STUDENT SOLUTIONS

Adeliza Flores
Las Positas College

GENERAL, ORGANIC, AND BIOLOGICAL CHEMISTRY: AN INTEGRATED APPROACH

Third Edition

Kenneth W. Raymond
Eastern Washington University

WILEY

JOHN WILEY & SONS, INC.

Cover image: © Wolfgang Kraus/iStockphoto

ISBN 978-0-470-55495-1

Printed in the United States of America

10 9 8 7 6 5 4 3 2 1

Printed and bound by Malloy Lithographers

Table Of Contents

Chapter 1
Science and Measurements

Solutions to Problems

1.1   *Which drawing correctly shows the relationship between pounds and kilograms?*

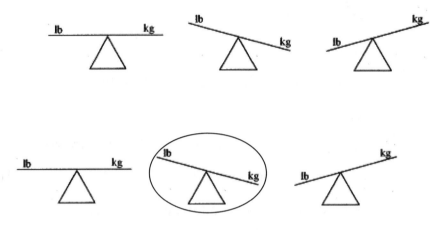

1 kilogram is heavier than 1 pound (1 kilogram = 2.205 lb).

1.3   *Is the statement "What goes up must come down" a scientific law or scientific theory? Explain.*

A law. It describes what is observed but does not explain why it happens.

1.5   *How is a theory different from a hypothesis?*

A hypothesis is a tentative explanation based on presently known facts while a theory is an experimentally tested explanation that is consistent with existing experimental evidence and accurately predicts the results of future experiments.

1.7   *Define the terms "matter" and "energy."*

Matter is anything that has mass and occupies space. Energy is the ability to do work or heat up something.

1.9   *On a hot day, a glass of iced tea is placed on a table.*

      a. What are some of the physical properties of the ice?

Ice is water in the solid state. It is clear and colorless, hard, and feels cold to the touch.

      b. What change in physical state would you expect to take place if the iced tea sits in the sun for a while?

The ice would slowly melt and turn into liquid water if the iced tea is left to sit in the sun for a while (melting).

1.11   *Give an example of a physical change that involves starting with a liquid and ending up with a gas.*

Some examples of physical changes are: a puddle of water evaporating from the ground, rubbing alcohol evaporating from the skin, spilled gasoline evaporating from the ground.

1.13   *What is potential energy?*

Potential energy is stored energy.

1.15   *Describe a situation where an object's potential energy varies as a result of changes in its position.*

As a Ferris wheel goes around, the potential energy of a rider changes relative to their position. When the rider is at the very top of the wheel, the potential energy is highest. As the wheel goes around, the potential energy of the rider decreases as they approach the bottom of the wheel. At the lowest point, the potential energy is lowest.

1.17   *A battery-powered remote control toy car sits at the bottom of a hill. The car begins to move and is steered up the hill.*

a. Describe the changes to the car's kinetic energy.

Kinetic energy is the energy associated with moving objects while potential energy is stored energy. Before the car starts to move up the hill, its kinetic energy is zero. As it begins its motion, the car gains kinetic energy.

b. Describe the changes to the car's potential energy that are related to its position.

At the bottom of the hill, the car has lower potential energy. As the car moves uphill, the car gains potential energy.

1.19   *Suppose that you are camping in the winter. To obtain drinking water, you use a propane-fueled camp stove to melt snow.*

a. Is the melting of snow a physical change? Explain.

Melting snow is a physical change because the chemical composition does not change.

b. When propane burns, the gases carbon dioxide and water vapor are formed. Is the burning of propane a physical change? Explain.

Burning propane is a chemical change because new substances are produced.

c. Describe the potential energy change that takes place for propane as it burns in the stove.

The potential energy stored in the propane is transformed into kinetic energy as the propane burns.

d. Describe the kinetic energy change that takes place for water as the snow melts.

As the snow melts, the water molecules move faster and so their kinetic energy increases.

1.21   *Based on your experience or the information in Table 1.1, which is larger?*

a. 1 yd or 1 m

1 m.   One meter is slightly larger than 1 yard because 1 yard is 3 feet and 1 meter is 3.281 feet. Therefore, 1 meter is 1.094 yard.

b. 1 lb or 1 g

1 lb. One kilogram is 2.205 pounds. Therefore one pound is 453.5 grams.

c. 1 cup or 1 mL

1 cup. One cup is 8 fluid ounces. One fluid ounce is 29.6 milliliter. Therefore, 1 cup is 237 mL.

1.23   *Based on your experience or the information in Table 1.1, which is larger?*

a. 1 mg or 1 μg?

1 mg. One mg equals 1000 μg.

b. 1 grain or 1 mg?

1 gr. One grain equals 65 mg.

c. 1 T or 1 tsp?

1 T. One T equals 15 mL which equals 3 tsp. Therefore, 1T = 3 tsp.

d. 1 T or 1 fl oz?

1fl oz. One fl oz equals 2T.

1.25   *Convert each number into scientific notation.*

a. 1,300		1.3×10^3
b. 6,901,000		6.901×10^6
c. 0.00013		1.3×10^{-4}
d. 0.0000006901		6.901×10^{-7}

1.27   Convert each number into ordinary notation.

    a. 7×10^{-2} 0.07

    b. 7×10^{2} 700

    c. 8.3×10^{8} 830,000,000

    d. 8.3×10^{-8} 0.000000083

1.29   a. In the year 2000, the world population is estimated to have been about 6×10^{9}. Convert this value into ordinary notation.

6,000,000,000

    b. In the year 1000, the world population is estimated to have been about 3×10^{6}. Convert this value into ordinary notation.

3,000,000

1.31   Express each value using an appropriate metric prefix.

    a. one-thousandth of a meter   1 millimeter

    b. one million meters   1 megameter

    c. one billion meters   1 gigameter

1.33   Express each distance in scientific notation and ordinary (decimal) notation, without using metric prefixes (example: 6.2 cm = 6.2×10^{-2} m = 0.062 m)

    a. 1.5 km   = 1.5×10^{3} m   = 1,500 m

    b. 5.67 mm   = 5.67×10^{-3} m   = 0.00567 m

    c. 5.67 nm   = 5.67×10^{-9} m   = 0.00000000567 m

    d. 0.3 cm   = 3×10^{-3} m   = 0.003 m

1.35   *Which is the greater amount of energy?*

a. 1 kcal or 1 kJ?

1 kcal. One calorie is larger than one joule (1 cal = 4.184 J), so 1 kcal (1000 cal) is larger than 1 kJ (1000 J).

b. 4.184 cal or 1 J?

4.184 cal. One calorie is larger than one joule (1 cal = 4.184 J), so 4.184 cal is larger than 1 J.

1.37   *a. How many meters are in 1 km?*

The prefix kilo (k) stands for 10^3. Therefore, 1 km = 1 x 10^3 m.

b. How many meters are in 5 km?

The prefix kilo (k) stands for 10^3. Therefore, 5 km = 5 x 10^3 m.

c. How many meters are in 10 km?

The prefix kilo (k) stands for 10^3. Therefore, 10 km = 10 x 10^3 m or 1 x 10^4 m.

1.39   *How many significant figures does each number have? Assume that each is a measured value.*

a. 1000000.5	All the zeroes count because they are between nonzero digits.
	8
b. 887.60	The ending zero counts because there is a decimal point.
	5

c. 0.668   Zeroes at the beginning of a number are not significant.

3

d. 45   All non-zero digits are significant.

2

e. 0.00045   Zeroes at the beginning of a number are not significant.

2

f. 70.   The ending zero counts because there is a decimal point.

2

1.41   *Solve each calculation, reporting each answer with the correct number of significant figures. Assume that each value is a measured value.*

a. 14 x 3.6

In multiplication and division, the answer is rounded to have the same number of significant figures as the measurement with the fewest significant figures. Since both measured values have two significant figures, the calculator answer (50.4) rounds to 50. or 5.0×10^1. The decimal point is necessary to indicate that the zero is significant.

5.0×10^1

b. 0.0027 ÷ 6.7784

The same rule applies as in part a. Dividing 0.0027 (2 significant figures) by 6.7784 (5 significant figures) gives 0.000398324, which rounds to 0.00040 or 4.0×10^{-4} (2 significant figures).

4.0×10^{-4}

c. 12.567 + 34

When adding or subtracting, the answer should have the same number of decimal places as the quantity with the fewest decimal places.

12. 567	Three decimal places
34	Zero decimal places

46.567   Rounds to 47 (0 decimal places)

47

d. *(1.2 x 10³ x 0.66) + 1.0*

When more than one operation is involved, calculate the part in parentheses first and round it to the appropriate significant figures and then perform the next part of the calculation. In the first calculation, both numbers have two significant figures, so the answer $(1.2 \times 10^3 \times 0.66 = 792)$ is rounded to two significant figures (790).

790	10s place is significant
1.0	10ths place is significant

791.0   Rounds to 790 with the 10s place significant

7.9×10^2

1.43   *A microbiologist wants to know the circumference of a cell being viewed through a microscope. Estimating the diameter of the cell to be 11 μm and knowing that circumference = π x diameter (we will assume that the cell is round, even though that is usually not the case), the microbiologist uses a calculator and gets the answer 34.55751919 μm. Taking significant figures into account, what answer should actually be reported? (π = 3.141592654.....).*

In the calculation π x diameter, the diameter (11 μm) has the fewest number of significant figures (2). The answer (34.55751414... μm) is reported with two significant figures.

35 μm

1.45   *You are at the state fair and pay a dollar for the chance to throw three baseballs in an attempt to knock over a pyramid of bowling pins. After your three tosses, the pins remain standing. Which of the following statements about your throws might be correct?*

    a. They were precise and accurate.
    b. They were neither precise nor accurate.
    c. They were precise but not accurate.

Either b or c could be correct since the pins remain standing. Your throws may or may not have been precise.

1.47   *Give the two conversion factors that are based on each equality.*

    a. 12 eggs = 1 dozen $\dfrac{12\ eggs}{1\ dozen}$ or $\dfrac{1\ dozen}{12\ eggs}$

    b. 1 x 10³ m = 1 km $\dfrac{1 \times 10^{3}\ m}{1\ km}$ or $\dfrac{1\ km}{1 \times 10^{3}\ m}$

    c. 0.946 L = 1 qt $\dfrac{0.946\ L}{1\ qt}$ or $\dfrac{1\ qt}{0.946\ L}$

1.49   Convert

    a. 48 eggs into dozen

12 eggs = 1 dozen

$$48\ \cancel{eggs}\ \times\ \frac{1\ dozen}{12\ \cancel{eggs}}\ =\ 4\ dozens$$

b. 250 m into kilometers

1×10^3 m = 1 km

$250 \ \cancel{m} \ \times \ \dfrac{1 \text{ km}}{1 \times 10^3 \ \cancel{m}} \ = 0.25 \text{ km}$

c. 2.7 L into quarts

0.946 L = 1 qt

$2.7 \ \cancel{L} \ \times \ \dfrac{1 \text{ qt}}{0.946 \ \cancel{L}} \ = 2.9 \text{ qt}$

1.51   *Convert*

a. 92 μg into grams.

$1 \ \mu g = 1 \times 10^{-6}$ g

$92 \ \cancel{\mu g} \ \times \ \dfrac{1 \times 10^{-6} \text{ g}}{1 \ \cancel{\mu g}} \ = 9.2 \times 10^{-5} \text{ g}$

b. 27.2 ng into milligrams.

$1 \text{ ng} = 1 \times 10^{-9}$ g; $1 \text{ mg} = 1 \times 10^{-3}$ g.

$27.2 \ \cancel{ng} \ \times \ \dfrac{1 \times 10^{-9} \ \cancel{g}}{1 \ \cancel{ng}} \ \times \ \dfrac{1 \text{ mg}}{1 \times 10^{-3} \ \cancel{g}} \ = 2.72 \times 10^{-5} \text{ mg}$

c. 0.33 kg into milligrams.

$1 \text{ kg} = 1 \times 10^3$ g; $1 \text{ mg} = 1 \times 10^{-3}$ g

$0.33 \ \cancel{kg} \ \times \ \dfrac{1 \times 10^3 \ \cancel{g}}{1 \ \cancel{kg}} \ \times \ \dfrac{1 \text{ mg}}{1 \times 10^{-3} \ \cancel{g}} \ = 3.3 \times 10^5 \text{ mg}$

d. 7.27 mg into micrograms.

$1 \text{ mg} = 1 \times 10^{-3} \text{ g}$; $1 \mu g = 1 \times 10^{-6} \text{ g}$

$$7.27 \text{ mg} \times \frac{1 \times 10^{-3} \text{ g}}{1 \text{ mg}} \times \frac{1 \mu g}{1 \times 10^{-6} \text{ g}} = 7.27 \times 10^{3} \mu g$$

1.53   Convert

a. 1 cm into kilometers

$1 \text{ cm} = 1 \times 10^{-2} \text{ m}$; $1 \text{ km} = 1 \times 10^{3} \text{ m}$

$$1 \text{ cm} \times \frac{1 \times 10^{-2} \text{ m}}{1 \text{ cm}} \times \frac{1 \text{ km}}{1 \times 10^{3} \text{ m}} = 1 \times 10^{-5} \text{ km}$$

b. 25 pm into micrometers

$1 \text{ pm} = 1 \times 10^{-12} \text{ m}$; $1 \mu m = 1 \times 10^{-6} \text{ m}$

$$25 \text{ pm} \times \frac{1 \times 10^{-12} \text{ m}}{1 \text{ pm}} \times \frac{1 \mu m}{1 \times 10^{-6} \text{ m}} = 2.5 \times 10^{-5} \mu m$$

c. 3.0 x 10^{-4} mm into decimeters

$1 \text{ mm} = 1 \times 10^{-3} \text{ m}$; $1 \text{ dm} = 1 \times 10^{-1} \text{ m}$

$$3.0 \times 10^{-4} \text{ mm} \times \frac{1 \times 10^{-3} \text{ m}}{1 \text{ mm}} \times \frac{1 \text{ dm}}{1 \times 10^{-1} \text{ m}} = 3.0 \times 10^{-6} \text{ dm}$$

d. 8.5 x 10^{-3} mm into nanometers

$1 \text{ mm} = 1 \times 10^{-3} \text{ m}$; $1 \text{ nm} = 1 \times 10^{-9} \text{ m}$

$$8.5 \times 10^{-3} \text{ mm} \times \frac{1 \times 10^{-3} \text{ m}}{1 \text{ mm}} \times \frac{1 \text{ nm}}{1 \times 10^{-9} \text{ m}} = 8.5 \times 10^{3} \text{ nm}$$

1.55   *Convert your weight from pounds to kilograms.*

Answers will vary depending on your weight. Below is a sample setup.

If your weight is 175 lb then

$$175 \text{ lb} \times \frac{1 \text{ kg}}{2.205 \text{ lb}} = 79.4 \text{ kg}$$

In general, answer: Your pound weight $\times \dfrac{1 \text{ kg}}{2.205 \text{ lb}}$.

1.57   Convert

a. 91°F into degrees Celsius

$$°C = \frac{°F - 32}{1.8} = \frac{91 - 32}{1.8} = 33°C$$

b. 53°C into degrees Fahrenheit

$$°F = (1.8 \times °C) + 32 = (1.8 \times 53°C) + 32 = 127°F$$

c. 0°C into kelvins

$$K = °C + 273 = 0°C + 273 = 273 \text{ K}$$

d. 309 K into degrees Celsius

$$°C = K - 273 = 309 \text{ K} - 273 = 36°C$$

1.59   *In 2008, 43,640 people completed the 12 km Bloomsday race in Spokane, WA. What is the distance of this race in miles?*

12 km = distance of the race

1 mi = 1.609 km

$$12 \text{ km} \times \frac{1 \text{ mi}}{1.609 \text{ km}} = 7.5 \text{ mi}$$

1.61   *Stavudine is an antiviral drug that has been tested as a treatment for AIDS. The daily recommended dosage of stavudine is 1.0 mg/kg of body weight. How many grams of this drug should be administered to a 150 lb patient?*

First, convert 150 lb to kg using the equivalence 2.205 lb = 1 kg. Then, use 1.0 mg/kg as a conversion factor to calculate the number of milligrams required by the patient. To convert the final answer to grams, use 1000 mg = 1 g.

$$150 \text{ lb} \times \frac{1 \text{ kg}}{2.205 \text{ lb}} \times \frac{1.0 \text{ mg}}{1 \text{ kg}} \times \frac{1 \text{ g}}{1000 \text{ mg}} = 0.068 \text{ g}$$

1.63   *Ivermectin is used to treat dogs that have intestinal parasites. The effective dosage of this drug is 10.5 μg/kg of body weight. How much ivermectin should be given to a 9.0 kg dog?*

$$9.0 \text{ kg} \times \frac{10.5 \text{ μg}}{1 \text{ kg}} = 95 \text{ μg or } 9.5 \times 10^{-5} \text{ g}$$

1.65   *The tranquilizer Valium is sold in 2.0 mL syringes that contain 50.0 mg of drug per 1.0 mL of liquid (50.0 mg/1.0 mL). If a physician prescribes 25 mg of this drug, how many milliliters should be administered?*

Note that the size of the syringes does not have anything to do with the solution of the problem, since it asks for milliliters needed and not the number of syringes.

$$25 \text{ mg} \times \frac{1.0 \text{ mL}}{50.0 \text{ mg}} = 0.50 \text{ mL}$$

1.67 *a. A vial contains 25 mg/mL of a particular drug. To administer 15 mg of the drug, how many milliliters should be drawn from the vial?*

Use the drug concentration in the vial 25 mg/mL (that is, 25 mg of the active drug is contained in every 1.0 mL of the vial content) as a conversion factor. To calculate how many milliliters of the drug should be dispensed to administer a dose of 15 mg:

$$15 \; \text{mg drug} \times \frac{1 \; \text{mL}}{25 \; \text{mg drug}} = 0.60 \; \text{mL}$$

b. A patient is to receive 50 cc of a drug mixture intravenously over a 1 hr time period. What is the appropriate IV drip rate in gtt/min?

This problem is multi-step and requires more than one conversion factor to complete. Use the conversion factor

15 drops (gtt) = 1 milliliter (mL).

To use this relationship, we first have to convert 50 cc to mL using the equality 1 cc = 1 mL. Because the problem asks for gtt/min, we also have to convert hour to minute.

Combining all of these steps results in the following calculation:

$$\frac{50 \; \text{cc}}{1 \; \text{hr}} \times \frac{1 \; \text{mL}}{1 \; \text{cc}} \times \frac{15 \; \text{gtt}}{1 \; \text{mL}} \times \frac{1 \; \text{hr}}{60 \; \text{min}} = 13 \; \text{gtt/min}$$

1.69 *At 20 °C, what is the mass in grams of (See Table 1.8)*

a. 2.0 mL of water?
b. 2.0 mL of whole blood?
c. 15.3 cm³ of salt?
d. 71.2 cm³ of lead?

To solve for the mass in grams, use the density as a conversion factor:

a. 2.0 mL of water?

$$2.0 \; \text{mL} \times \frac{1.00 \; \text{g}}{1 \; \text{mL}} = 2.0 \; \text{g water}$$

14

b. *2.0 mL of whole blood?*

$$2.0 \;\cancel{mL} \;\times\; \frac{1.06 \text{ g}}{1 \;\cancel{mL}} \;=\; 2.1 \text{ g whole blood}$$

c. *15.3 cm³ of salt?*

$$15.3 \;\cancel{cm^3} \;\times\; \frac{2.17 \text{ g}}{1 \;\cancel{cm^3}} \;=\; 33.2 \text{ g salt}$$

d. *71.2 cm³ of lead?*

$$71.2 \;\cancel{cm^3} \;\times\; \frac{11.35 \text{ g}}{1 \;\cancel{cm^3}} \;=\; 808 \text{ g lead}$$

1.71 *At 20 °C what is the volume in milliliters occupied by (See Table 1.8)*

a. *15.2 g of water?*
b. *2.0 kg of kerosene?*
c. *9.2 x 10⁻² g of isopropyl alcohol?*
d. *75 g chloroform?*

To solve for the volume in milliliters, use the density as a conversion factor:

a. *15.2 g of water?*

$$15.2 \;\cancel{g} \;\times\; \frac{1 \text{ mL}}{1.00 \;\cancel{g}} \;=\; 15.2 \text{ mL water}$$

b. *2.0 kg of kerosene?*

$$2.0 \;\cancel{kg} \;\times\; \frac{1 \times 10^3 \;\cancel{g}}{1 \;\cancel{kg}} \;\times\; \frac{1 \text{ mL}}{0.82 \;\cancel{g}} \;=\; 2.4 \times 10^3 \text{ mL kerosene}$$

c. *9.2 x 10⁻² g of isopropyl alcohol?*

$$9.2 \times 10^{-2} \;\cancel{g} \;\times\; \frac{1 \text{ mL}}{0.785 \;\cancel{g}} \;=\; 0.12 \text{ mL isopropyl alcohol}$$

d. 75 g chloroform?

$$75 \text{ g} \times \frac{1 \text{ mL}}{1.49 \text{ g}} = 5.0 \times 10^1 \text{ mL chloroform}$$

1.73 *A patient has 25.0 mL of blood drawn and this volume of blood has a mass of 26.5 g. What is the density of the blood?*

Density is expressed in g/mL. Therefore, the density is found by dividing the mass of the blood by its volume.

$$\text{Density} = \frac{\text{mass}}{\text{volume}} = \frac{26.5 \text{ g}}{25.0 \text{ mL}} = 1.06 \text{ g/mL}$$

1.75 *What is the specific gravity of whole blood at 20°C? (See Table 1.8.)*

The specific gravity relates the density of a substance to that of water at the same temperature:

$$\text{specific gravity} = \frac{\text{density of substance}}{\text{density of water}}$$

At 20°C, the density of water is 1.00 g/mL and the density of whole blood is 1.06 g/mL.

$$\text{specific gravity of whole blood} = \frac{1.06 \text{ g/mL}}{1.00 \text{ g/mL}} = 1.06$$

1.77 *Calculate the number of calories of heat energy required for each (See Table 1.9)*

a. to warm 35.0 g of water from 21.0 °C to 29.0 °C

Use the specific heat of water, 1.000 cal/g°C (Table 1.9), to convert mass and temperature change into calories. The temperature change is 8.0 °C (29.0 °C– 21.0 °C)

$$35.0 \text{ g} \times 8.0 \text{ °C} \times \frac{1.000 \text{ cal}}{\text{g °C}} = 280 \text{ cal}$$

b. to warm 17.5 g of water from 18.0 °C to 54.0 °C

$$17.5 \; \cancel{g} \; \times \; 36.0 \; \cancel{°C} \; \times \; \frac{1.000 \; cal}{\cancel{g} \; \cancel{°C}} \; = \; 6.30 \times 10^2 \; cal$$

1.79   *Calculate the number of calories of heat energy required for each (See Table 1.9)*

a. to warm 35.0 mL of water from 21.0 °C to 29.0 °C

First, use the density of water to convert mL of water to g of water.

$$35.0 \; \cancel{mL} \; \times \; \frac{1.00 \; g}{1 \; \cancel{mL}} \; = \; 35.0 \; g \; water$$

Then, use the specific heat of water, 1.000 cal/g°C (Table 1.9), to convert mass and temperature change into calories. The temperature change is 8.0 °C (29.0 °C–21.0 °C)

$$35.0 \; \cancel{g} \; \times \; 8.0 \; \cancel{°C} \; \times \; \frac{1.000 \; cal}{\cancel{g} \; \cancel{°C}} \; = \; 280 \; cal$$

b. to warm 17.5 mL of water from 18.0 °C to 54.0 °C

$$17.5 \; \cancel{mL} \; \times \; \frac{1.00 \; g}{1 \; \cancel{mL}} \; = \; 17.5 \; g \; water$$

$$17.5 \; \cancel{g} \; \times \; 36.0 \; \cancel{°C} \; \times \; \frac{1.000 \; cal}{\cancel{g} \; \cancel{°C}} \; = \; 6.30 \times 10^2 \; cal$$

1.81   *How much will the temperature change when 750 g of each of the following materials absorbs 1.25 x 10⁴ cal of heat energy?*

To do these calculations, substitute all the known values into the conversion equation using specific heat and solve for the unknown (temperature change) *a. iron (specific heat = 0.11 cal/g°C)*

$$750 \text{ g} \times \text{temperature change} \times \frac{0.11 \text{ cal}}{\text{g °C}} = 1.25 \times 10^4 \text{ cal}$$

$$\text{temperature change} = \frac{1.25 \times 10^4 \text{ cal}}{750 \text{ g}} \times \frac{\text{g °C}}{0.11 \text{ cal}} = 150 \text{ °C}$$

b. stainless steel (specific heat = 0.12 cal/g°C)

$$750 \text{ g} \times \text{temperature change} \times \frac{0.12 \text{ cal}}{\text{g °C}} = 1.25 \times 10^4 \text{ cal}$$

$$\text{temperature change} = \frac{1.25 \times 10^4 \text{ cal}}{750 \text{ g}} \times \frac{\text{g °C}}{0.12 \text{ cal}} = 140 \text{ °C}$$

c. aluminum (specific heat = 0.89 J/g°C)

First, convert 1.25×10^4 cal to J using the relationship given in Table 1.1.

$$1.25 \times 10^4 \text{ cal} \times \frac{4.184 \text{ J}}{1 \text{ cal}} = 52,300 \text{ J}$$

$$750 \text{ g} \times \text{temperature change} \times \frac{0.89 \text{ J}}{\text{g °C}} = 52,300 \text{ J}$$

$$\text{temperature change} = \frac{52,300 \text{ J}}{750 \text{ g}} \times \frac{\text{g °C}}{0.89 \text{ J}} = 78 \text{ °C}$$

1.83 *In the past 200 years, in what ways have scientific discoveries led to changes in the treatment of diabetes?*

Insulin became available, the purity of insulin was improved, genetically engineered human insulin was put on the market, and oral drugs were developed.

1.87   a. *A 6' 2" tall adult weighs 180 lbs. What is his BMI? Based on this value, what is his status: underweight, normal, overweight, or obese?*

The formula for BMI is given below, where weight is given in pounds and height is given in inches. First, calculate the total inches in height.

$$BMI = 703 \times \frac{weight}{height^2}$$

$$height = 6\ ft\ 2\ in = 6\ \cancel{ft} \times \frac{12\ in}{1\ \cancel{ft}} + 2\ in = 74\ in$$

$$BMI = 703 \times \frac{180\ lbs}{(74)^2}$$

$$BMI = 23$$

According to Table 1.7, this falls within the "recommended weight" status.

b. *Answer part a, but using your height and weight.*

Answers will vary.

c. *A woman stands 1.65 m tall and weighs 72.7 kg. What is her BMI and what is her status?*

First, convert m to in and then kg to lb. Then use the same formula as above.

$$1.65\ \cancel{m} \times \frac{3.281\ \cancel{ft}}{1\ \cancel{m}} \times \frac{12\ in}{1\ \cancel{ft}} = 65.0\ in$$

$$72.7\ \cancel{kg} \times \frac{2.205\ \cancel{lb}}{1\ \cancel{kg}} = 1.60 \times 10^2\ lb$$

$$BMI = 703 \times \frac{weight}{height^2}$$

$$BMI = 703 \times \frac{1.60 \times 10^2\ lbs}{(65.0)^2}$$

$$BMI = 26.6$$

This BMI places her in the overweight status.

1.89   *A patient has a temperature of 31°C. Should her clinician be concerned?*

   If a patient has a temperature of 31°C, her physician should be concerned because this is considerably lower than average normal temperature (37°C), even after factoring temperature measurement errors (up to 2°C) due to variation in body location from which measurements are made.

1.91   *One of the rule changes that the NCAA made to discourage rapid weight loss was to shorten the time between weigh in and competition from 24 hours to just 2 hours. Why would this discourage athletes from trying to make weight?*

   Two hours might not be enough time to rehydrate and to recover from the effects of dehydration.

1.93   *a. Use the density of water (1.00 g/mL) to derive a conversion factor for water that has the units lb/cup.*

$$\frac{1.00 \ \cancel{g}}{1.00 \ \cancel{mL}} \ \text{x} \ \frac{1 \ \text{lb}}{454 \ \cancel{g}} \ \text{x} \ \frac{237 \ \cancel{mL}}{1 \ \text{cup}} = \frac{0.522 \ \text{lb}}{1 \ \text{cup}}$$

   b. If an athlete reduces her body's water volume by 5.5 cups through restricting fluid intake and sweating in a sauna, how much weight has she lost? Is this a good idea? Explain.

$$5.5 \ \cancel{cup} \ \text{x} \ \frac{0.522 \ \text{lb}}{1 \ \cancel{cup}} = 2.9 \ \text{lbs}$$

   No. Losing weight through dehydration can adversely affect endurance, strength, energy, and motivation. Extreme dehydration can lead to kidney failure, heart attack, and death.

1.95   *a.* Write the two conversion factors that are based on the equality 1 grain = 325 milligrams.

$$\frac{1 \ \text{grain}}{325 \ \text{mg}} \ \text{and} \ \frac{325 \ \text{mg}}{1 \ \text{grain}}$$

b. Which conversion factory in your answer to part a would be used to convert grains to milligrams?

$$\frac{325 \text{ mg}}{1 \text{ grain}}$$

c. One aspirin tablet contains 5.0 grains of aspirin. How many milligrams of aspirin are in two tablets?

$$2 \cancel{\text{ tablets }} \times \frac{5.0 \cancel{\text{ grains}}}{1 \cancel{\text{ tablet}}} \times \frac{325 \text{ mg}}{1 \cancel{\text{ grain}}} = 3300 \text{ mg}$$

d. How many grams of aspirin are in two tablets?

$$3300 \cancel{\text{ mg}} \times \frac{1 \times 10^{-3} \text{ g}}{1 \cancel{\text{ mg}}} = 3.3 \text{ g}$$

e. How many micrograms of aspirin are in two tablets?

$$3.3 \cancel{\text{ g}} \times \frac{1 \text{ µg}}{1 \times 10^{-6} \cancel{\text{ g}}} = 3.3 \times 10^6 \text{ µg}$$

1.97   *Suppose that you were on a soccer team and the coach asked you to swallow a pill-like temperature sensor. What would your reaction be?*

A response to this question is a matter of opinion and may vary from student to student.

Chapter 2
Atoms and Elements

Solutions to Problems

2.1 *Identify the two nuclei and the type of nuclear radiation that this released.*

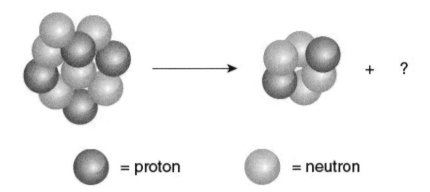

To identify the two nuclei, use the number of protons, which is the same as the atomic number. Refer to the periodic table to determine the element.

The first nucleus belongs to a Be atom because it contains 4 protons. Its mass number is 10 (4 protons + 6 neutrons).

The product nucleus belongs to a He atom because it contains 2 protons. Its mass number is 6 (2 protons + 4 neutrons).

The difference between the two nuclei corresponds to a loss of 2 protons and 2 neutrons released as an alpha particle radiation. The equation is:

$$^{10}_{4}Be \ \rightarrow \ ^{6}_{2}He \ + \ ^{4}_{2}\alpha$$

2.3 *Why does the nucleus of an atom have a positive charge?*

It consists of protons (positive charge) and neutrons (no charge).

2.5 *a. A hydrogen atom has a diameter of 5.3 x 10^{-11} m. Express this diameter in nanometers and picometers (pico=10^{-12}).*
b. Placed side-by-side, how many hydrogen atoms would it take to equal a distance of 1 inch (1 in = 2.54 cm)?

Set up the following conversion steps to solve these two problems.

a. *A hydrogen atom has a diameter of 5.3 x 10^{-11} m. Express this diameter in nanometers and picometers (pico=10^{-12}).*

$$5.3 \times 10^{-11} \; \cancel{m} \; \times \; \frac{1 \; nm}{1 \times 10^{-9} \; \cancel{m}} \; = \; 5.3 \times 10^{-2} \; nm \; or \; 0.053 \; nm$$

$$5.3 \times 10^{-11} \; \cancel{m} \; \times \; \frac{1 \; pm}{1 \times 10^{-12} \; \cancel{m}} \; = \; 5.3 \times 10^{1} \; pm \; or \; 53 \; pm$$

b. *Placed side-by-side, how many hydrogen atoms would it take to equal a distance of 1 inch (1 in = 2.54 cm)?*

First, convert the diameter of one hydrogen atom to cm:

$$5.3 \times 10^{-11} \; \cancel{m} \; \times \; \frac{1 \; cm}{1 \times 10^{-2} \; \cancel{m}} \; = \; 5.3 \times 10^{-9} \; cm$$

Then, set up a conversion to determine how many hydrogen atoms would "fit" into a distance of 1 inch or 2.54 cm using the equivalence of 1 hydrogen atom occupying 5.3 x 10^{-9} cm:

$$2.54 \; \cancel{cm} \; \times \; \frac{1 \; hydrogen \; atom}{5.3 \times 10^{-9} \; \cancel{cm}} \; = \; 4.8 \times 10^{8} \; hydrogen \; atoms$$

2.7   *A tin atom has a diameter of 145 pm (p = pico = 10^{-12}). If the nucleus of an atom is 100,000 times smaller than the atom, what is the diameter of the nucleus of a tin atom in picometers?*

According to the problem, the diameter of the nucleus of a tin atom is 1/100,000 that of the diameter of the whole atom:

$$\frac{1}{100,000} \; \times \; 145 \; pm \; = \; 1.45 \times 10^{-3} \; pm$$

2.9   *Give the names and atomic symbols for the four most abundant elements in the universe.*

hydrogen (H), helium (He), oxygen (O), nitrogen (N)

2.11   *Give the names and atomic symbols for the four most abundant elements in the human body.*

oxygen (O), carbon (C), hydrogen (H), nitrogen (N)

2.13   *The elements carbon, calcium, chlorine, copper, and cobalt are present in the human body. Match each element to its correct atomic symbol:*
Co, Cu, Ca, C, Cl.

Co is cobalt; **Cu** is copper; **Ca** is calcium; **C** is carbon; **Cl** is chlorine

2.15   *How many protons and neutrons are present in the nucleus of each?*

The number of protons is equal to the atomic number. The number of neutrons is equal to the mass number minus the number of protons.

a. $^{19}_{9}F$ 9 protons   19 - 9 = 10 neutrons

b. $^{23}_{11}Na$ 11 protons   23 - 11 = 12 neutrons

c. $^{238}_{92}U$ 92 protons   238 - 92 = 146 neutrons

2.17   *Give the atomic number and mass number of*
a. a helium atom with 2 neutrons
b. a lithium atom with 3 neutrons
c. a neon atom with 10 neutrons

Use the periodic table to look up the atomic number for each element. The mass number can be calculated by adding the number of protons (equal to the atomic number) and the number of neutrons.

		Atomic number	Mass number
a.	*a helium atom with 2 neutrons*	2	2 protons + 2 neutrons = 4
b.	*a lithium atom with 3 neutrons*	3	3 protons + 3 neutrons = 6
c.	*a neon atom with 10 neutrons*	10	10 protons + 10 neutrons = 20

2.19   *Give the atomic notation for each atom described in Problem 2.17*

The general symbol for atomic notation is $_Z^A X$, where A is the mass number, Z is the atomic number, and X is the symbol of the element.

a. $_2^4 He$ b. $_3^6 Li$ c. $_{10}^{20} Ne$

2.21   *Give the number of neutrons in an atom of*
a. $_{18}^{40} Ar$
b. $_{20}^{42} Ca$
c. $_{23}^{50} V$

To calculate the number of neutrons, subtract the atomic number from the mass number:

a. $_{18}^{40} Ar$ number of neutrons = 40 − 18 = 22 neutrons

b. $_{20}^{42} Ca$ number of neutrons = 42 − 20 = 22 neutrons

c. $_{23}^{50} V$ number of neutrons = 50 − 23 = 27 neutrons

2.23   *Give the number of electrons present in a neutral atom of each element.*
a. copper
b. calcium
c. chlorine

For a neutral atom of an element:

number of electrons = number of protons (equal to the atomic number)

a. copper   Cu   29 electrons

b. calcium   Ca   20 electrons

c. chlorine   Cl   17 electrons

2.25   *Complete the table.*

Answers are in **bold**:

Name	Calcium	**Carbon**	Copper
Atomic notation	$^{40}_{20}\textbf{Ca}$	$^{13}_{6}C$	$^{63}_{29}\textbf{Cu}$
Number of protons	**20**	6	29
Number of neutrons	**20**	7	34
Atomic number	20	**6**	29
Mass number	40	**13**	63

2.27   *Which of the following statements do not accurately describe isotopes of an element?*
a. same number of protons
b. same mass number
c. same atomic number

Choice *b*, the same mass number, would not accurately describe isotopes of an element.

Isotopes are atoms of the same element that have the same number of protons but different number of neutrons. Since the number of protons and atomic number are the same thing, then answers *a* and *c* are correct. The mass number would not be the same for isotopes of an element.

2.29   *Which are isotopes?*
$^{64}_{28}Ni$, $^{64}_{30}Zn$, $^{65}_{29}Cu$, $^{65}_{30}Zn$

Isotopes are atoms of the same element that have different numbers of neutrons.

$^{64}_{30}Zn$ and $^{65}_{30}Zn$ are isotopes.

2.31   *Rubidium, which appears in nature as $_{37}^{85}Rb$ (mass = 84.91 amu) and $_{37}^{87}Rb$ (mass = 86.91 amu) has an atomic weight of 85.47 amu. Which isotope predominates?*

$_{37}^{85}Rb$. The mass of this isotope (84.91 amu) is closest to the atomic weight given in the periodic table (85.47 amu) and therefore is the more abundant one.

2.33   *List some of the physical properties of metals.*

Good conductors of heat and electricity, shiny, malleable, and ductile.

2.35   *Identify two elements that belong to each group.*

a. halogen

Possible answers are: fluorine, chlorine, bromine, and iodine.

b. inert gas

Possible answers are: helium, neon, argon, krypton, xenon, and radon.

2.37   *Identify two representative nonmetallic elements that are in the third period.*

Possible answers are: phosphorus, sulfur, chlorine, and argon.

2.39   *Which of the two elements is more metallic?*

Refer to the periodic table and find where the elements are located. The farther they are to the left of the metal-nonmetal transition line ("zigzag" line), the more metallic they are. Elements to the right of this line are nonmetals. Another way of thinking of this is that the farther left in a period or the farther down a group, the more metallic the element is.

a. Na and Cl

Na. Na is found in the leftmost column of the periodic table and is more metallic than Cl (a nonmetal) which is to the right of the transition line.

b. O and Te

Te. Te is found closer to the right of the transition line than O and therefore is more metallic.

2.41   *Which would you expect to be larger, an atom of cesium or an atom of francium?*

Francium. Francium is below cesium in Group 1A of the periodic table. Generally, atomic size increases moving down a group.

2.43   *List the two conversion factors that relate number of lithium atoms and moles of lithium.*

Given that 1 mole of atoms is equal to 6.02×10^{23} atoms, the two conversion factors are:

$$\frac{6.02 \times 10^{23} \text{ Li atoms}}{1 \text{ mol Li}} \text{ and } \frac{1 \text{ mol Li}}{6.02 \times 10^{23} \text{ Li atoms}}$$

2.45   *a. How many moles of lithium is 3.01×10^{23} lithium atoms? (Refer to Problem 2.43)*

To convert number of Li atoms to mol Li atoms, use the conversion factor that has mol Li atoms in the numerator:

$$3.01 \times 10^{23} \text{ } \cancel{\text{Li atoms}} \text{ x } \frac{1 \text{ mol Li}}{6.02 \times 10^{23} \text{ } \cancel{\text{Li atoms}}} = 0.500 \text{ mol Li}$$

b. How many lithium atoms is 0.525 mol of lithium?

To convert mol Li atoms to number of Li atoms, use the conversion factor that has number of Li atoms in the numerator:

$$0.525 \text{ } \cancel{\text{mol Li}} \text{ x } \frac{6.02 \times 10^{23} \text{ Li atoms}}{1 \text{ } \cancel{\text{mol Li}}} = 3.16 \times 10^{23} \text{ Li atoms}$$

2.47   *List the two conversion factors that relate grams of calcium and moles of calcium.*

Use the molar mass of Ca which is numerically equal to its atomic weight in amu to set up the two conversion factors. From the periodic table, 1 mole of Ca weighs 40.08 g.

$$\frac{40.08 \text{ g Ca}}{1 \text{ mol Ca}} \quad \text{and} \quad \frac{1 \text{ mol Ca}}{40.08 \text{ g Ca}}$$

2.49   *a. How many moles of calcium is 126 g of Ca? (Refer to Problem 2.47)*

To convert grams of Ca to mol Ca, use the conversion factor that has mol Ca in the numerator:

$$126 \text{ g Ca} \times \frac{1 \text{ mol Ca}}{40.08 \text{ g Ca}} = 3.14 \text{ mol Ca}$$

b. How many grams of calcium is 2.25 mol of Ca?

To convert mol Ca to grams of Ca, use the conversion factor that has grams Ca in the numerator:

$$2.25 \text{ mol Ca} \times \frac{40.08 \text{ g Ca}}{1 \text{ mol Ca}} = 90.2 \text{ g Ca}$$

2.51   *How many atoms are present in 2.00 mol of aluminum?*

One mole of aluminum contains 6.02×10^{23} atoms.

$$2.00 \text{ mol Al} \times \frac{6.02 \times 10^{23} \text{ Al atoms}}{1 \text{ mol Al}} = 1.20 \times 10^{24} \text{ Al atoms}$$

2.53   *7.25×10^{12} sodium atoms is how many moles?*

One mole of sodium contains 6.02×10^{23} atoms.

$$7.25 \times 10^{12} \text{ Na atoms} \times \frac{1 \text{ mol Na}}{6.02 \times 10^{23} \text{ Na atoms}} = 1.20 \times 10^{-11} \text{ mol Na}$$

2.55   *What is the*

a. atomic weight of helium (He)?

4.00 amu (rounded to two decimal places)

b. molar mass of helium?

4.00 g/mol
The molar mass of an element is numerically equal to its atomic weight but expressed in g/mol.

c. mass (in grams) of 5.00 mol of helium?

One mole of He has a mass of 4.00 g.

$$5.00 \; \cancel{\text{mol He}} \; \times \; \frac{4.00 \text{ g He}}{1 \; \cancel{\text{mol He}}} = 20.0 \text{ g He}$$

d. number of helium atoms in 8.85x10^{-5} mol of helium?

One mole of He contains 6.02 x 10^{23} He atoms.

$$8.85 \times 10^{-5} \; \cancel{\text{mol He}} \; \times \; \frac{6.02 \times 10^{23} \text{ He atoms}}{1 \; \cancel{\text{mol He}}} = 5.33 \times 10^{19} \text{ He atoms}$$

e. mass (in grams) of 3.39 x 10^{20} helium atoms?

One mole of He has a mass of 4.00 g and contains 6.02 x 10^{23}He atoms.

$$3.39 \times 10^{20} \; \cancel{\text{He atoms}} \; \times \; \frac{1 \; \cancel{\text{mol He}}}{6.02 \times 10^{23} \; \cancel{\text{He atoms}}} \; \times \; \frac{4.00 \text{ g He}}{1 \; \cancel{\text{mol He}}} = 2.25 \times 10^{-3} \text{ g He}$$

2.57   *In 3.45 mg of boron there are how many of the following?*

a. grams of boron $3.45 \; \cancel{\text{mg}} \; \times \; \dfrac{1 \times 10^{-3} \text{ g}}{1 \; \cancel{\text{mg}}} = 3.45 \times 10^{-3} \text{ g B}$

b. moles of boron $3.45 \times 10^{-3} \; \cancel{\text{g B}} \; \times \; \dfrac{1 \text{ mol B}}{10.81 \; \cancel{\text{g B}}} = 3.19 \times 10^{-4} \text{ mol B}$

c. boron atoms

$$3.19 \times 10^{-4} \; \text{mol B} \; \times \; \frac{6.02 \times 10^{23} \; \text{B atoms}}{1 \; \text{mol B}} = 1.92 \times 10^{20} \; \text{B atoms}$$

2.59   *Helium has a different emission spectrum than hydrogen. Account for this difference.*

Helium's ground state electron arrangement is different from that of hydrogen.

2.61   *Using a periodic table, determine the number of electrons held in energy levels 1-3 of each atom.*

The electrons in an atom are arranged according to energy levels. Each energy level is specified by a number n. The maximum number of electrons that each energy level can hold is $2n^2$. When assigning the electrons of a given element to these energy levels, we start filling the $n=1$ level completely first and make our way up until all of the electrons have been assigned to an energy level.

	Total # of electrons	$n = 1$ (max. 2 e⁻)	$n = 2$ (max. 8 e⁻)	$n = 3$ (max. 18 e⁻)
a. Be	4	2	2	0
b. N	7	2	5	0
c. Cl	17	2	8	7

2.63   *Specify the number of valence electrons for each atom.*

Valence electrons are the electrons found in the outermost shell of an atom. The valence shell is the highest numbered, occupied energy level in an atom. Atoms in the same group have the same number of valence electrons. For representative elements, the number of valence electrons corresponds to the group number to which the element belongs.

a. H   1 valence electron
b. Be   2 valence electrons
c. C   4 valence electrons
d. Br   7 valence electrons
e. Ne   8 valence electrons

2.65   *For each, give the total number of electrons, the number of valence electrons, and the number of the energy level that holds the valence electrons.*

The energy level is correlated with the period number for the element.

	Total # of electrons	*Number of valence electrons*	*Energy level of valence electrons*
a. He	*2*	*2*	*1*
b. Xe	*54*	*8*	*5*
c. Te	*52*	*6*	*5*
d. Pb	*82*	*4*	*6*

2.67   *For a neutral atom of element 112*

a. how many total electrons are present?

112. In a neutral atom, the number of electrons is equal to the number of protons which is the same as the atomic number of the element.

b. how many valence electrons are present?

2. Element 112 is predicted to have 2 valence electrons like the element zinc which belongs to the same group.

c. which energy level holds the valence electrons?

Energy level $n = 7$. The valence electrons are held in the highest occupied energy level which is the same as the period number for that element. Element 112 belongs to Period 7 of the periodic table.

d. is the valence energy level full?

No, the $n = 7$ energy level can contain 98 electrons. There are only 2 electrons in the $n = 7$ energy level of element 112.

2.69   *Draw the electron dot structure of each atom.*

The electron dot structure shows the number of valence electrons, drawn as dots around the symbol for the element.

a. Na   Na•   *c. Ar*   :Är:

b. Cl    :C̈l•   *d. S*   •S̈•

2.71   *a. What is an alpha particle?*

An alpha particle is a type of radiation that can be emitted by radioisotopes.

b. How is an alpha particle similar to a helium nucleus?

Like a helium nucleus, an alpha particle has 2 protons, 2 neutrons, and 2+ charge.

c. How is an alpha particle different than a helium nucleus?

The alpha particle has a much greater energy than a helium nucleus.

2.73   *a. What is a positron?*

A positron is a subatomic particle that can be emitted by radioisotopes as radiation.

b. How is a positron similar to a beta particle?

Positrons have the same mass as beta particles and are emitted at speeds of up to 90% of the speed of light like beta particles.

c. How is a positron different than a beta particle?

A positron has a 1+ charge but a beta particle has 1- charge.

2.75   *Identify the missing product in each nuclear equation.*

In balancing nuclear equations, the sum of the mass numbers and the sum of the charges on atomic nuclei and subatomic particles must be the same on both sides of the equation.

a. $^{32}_{15}P \rightarrow ? + ^{0}_{1}\beta^{+}$

$^{32}_{15}P \rightarrow ^{32}_{14}Si + ^{0}_{1}\beta^{+}$

b. $^{40}_{19}K \rightarrow ? + ^{0}_{-1}\beta$

$^{40}_{19}K \rightarrow ^{40}_{20}Ca + ^{0}_{-1}\beta$

c. $^{40}_{19}K \rightarrow ^{40}_{18}Ar + ?$

$^{40}_{19}K \rightarrow ^{40}_{18}Ar + ^{0}_{1}\beta^{+}$

2.77   *a. Write the balanced nuclear equation for the loss of an alpha particle from* $^{203}_{83}Bi$.

The loss of an alpha particle results in the loss of 2 neutrons and 2 protons:

$$^{203}_{83}Bi \rightarrow ^{199}_{81}Tl + ^{4}_{2}\alpha$$

b. Write the balanced nuclear equation for the loss of a positron from $^{17}_{9}F$.

The loss of a positron results in the loss of 1 proton but no change in the mass number:

$$^{17}_{9}F \rightarrow ^{17}_{8}O + ^{0}_{1}\beta^{+}$$

2.79   *Write a balanced nuclear equation for each process.*

To find the atomic number of an isotope produced in the nuclear reaction, the atomic number of the emitted particle is subtracted from the atomic number of the original radioisotope. Looking this number up on the periodic table gives the atomic symbol of the newly produced isotope. The mass number of the emitted particle is subtracted from the mass number of the original radioisotope to give the new mass number.

a. $^{187}_{80}Hg$ *emits an alpha particle*

$$^{187}_{80}Hg \rightarrow ^{183}_{78}Pt + ^{4}_{2}\alpha$$

b. $^{266}_{88}$Ra *emits an alpha particle*

$$^{266}_{88}\text{Ra} \quad \rightarrow \quad ^{222}_{86}\text{Rn} \quad + \quad ^{4}_{2}\alpha$$

c. $^{238}_{92}$U *emits an alpha particle*

$$^{238}_{92}\text{U} \quad \rightarrow \quad ^{234}_{90}\text{Th} \quad + \quad ^{4}_{2}\alpha$$

2.81 $^{197}_{80}$*Hg, a beta and gamma emitter, is used for brain scans. Write a balanced equation for the loss of 1 beta particle and 1 gamma ray from this radioisotope.*

The loss of a beta particle results in the loss of 1 neutron and the gain of 1 proton. The loss of a gamma ray results in no change in the mass number and charge of the original isotope.

$$^{197}_{80}\text{Hg} \quad \rightarrow \quad ^{197}_{81}\text{Tl} \quad + \quad ^{0}_{-1}\beta \quad + \quad ^{0}_{0}\gamma$$

2.83   *Write the balanced nuclear equation for the loss of a positron from iodine-128.*

$$^{128}_{53}\text{I} \quad \rightarrow \quad ^{128}_{52}\text{Te} \quad + \quad ^{0}_{1}\beta^{+}$$

2.85   *In a radioactive decay series, one radioisotope decays into another, which decays into another, and so on. For example, in fourteen steps uranium-238 is converted to lead-206. Starting with uranium-238, the first decay in this series releases an alpha particle, the second decay releases a beta particle, and the third releases a beta particle. Write balanced nuclear equations for these three reactions.*

$$^{238}_{92}\text{U} \quad \rightarrow \quad ^{234}_{90}\text{Th} \quad + \quad ^{4}_{2}\alpha$$

$$^{234}_{90}\text{Th} \quad \rightarrow \quad ^{234}_{91}\text{Pa} \quad + \quad ^{0}_{-1}\beta$$

$$^{234}_{91}\text{Pa} \quad \rightarrow \quad ^{234}_{92}\text{U} \quad + \quad ^{0}_{-1}\beta$$

2.87 *Smoke detectors contain an alpha emitter. Considering the type of radiation released and the usual placement of a smoke detector, do these detectors pose a radiation risk? Explain.*

No. Since alpha particles travel only 4 -5 cm in air and smoke detectors are usually on the ceiling, the risk of exposure to alpha radiation is small.

2.89 *Radioisotopes used for diagnosis are beta, gamma, or positron emitters. Why are alpha emitters not used for diagnostic purposes?*

Alpha particles are relatively large and do not penetrate tissue very deeply. Radiation emissions must be able to pass through the body and reach a detector to be useful for diagnostic purposes.

2.91 $^{59}_{26}Fe$, *a beta emitter with a half-life of 45 days, is used to monitor iron metabolism.*

    a. Write a balanced nuclear equation for this radioactive decay.

$$^{59}_{26}\text{Fe} \rightarrow ^{59}_{27}\text{Co} + ^{0}_{-1}\beta$$

    b. How much time must elapse before a patient contains just 25% of an administered dose of $^{59}_{26}Fe$, assuming that this radioisotope is eliminated from the body only by radioactive decay?

90 days. The half-life of the iron radioisotope is 45 days. After 45 days, only 50% of the administered dose is left in the patient. After another 45 days, only half of the 50% remaining dose is left. Half of the 50% dose left is 25% of the original. Therefore, it takes 90 days for the amount of administered dose to go down to 25% of the original.

2.93 *Phosphorus-32 has a half-life of 14.3 days. Beginning with a 2.00 μg sample of this radioisotope,*

    a. how many micrograms remain after 2 half-lives?

Original amount	2.00 μg
After 1 half-life (14.3 days)	1.00 μg remain
After 2 half-lives (28.6 days)	**0.500 μg remain**

b. how many micrograms remain after 42.9 days?

$$\text{number of half-lives} \;=\; 42.9 \text{ days } \times \; \frac{1 \text{ half-life}}{14.3 \text{ days}} \;=\; 3 \text{ half-lives}$$

Continuing the series from part *a*:

Original amount	2.00 μg
After 1 half-life (14.3 days)	1.00 μg remain
After 2 half-lives (28.6 days)	0.500 μg remain
After 3 half-lives (42.9 days)	**0.250 μg remain**

c. how many days will it take for the mass of phosphorus-32 to drop to 0.125 μg?

57.2 days
0.125 μg is what remains after 4 half-lives (42.9 days + 14.3 days = 57.2 days).
Refer to the table above.

2.95   *a. Neon-19 is a positron emitter with a half-life of about 20 seconds. Write a balanced nuclear equation for the loss of a positron from this radioisotope.*

$$^{19}_{10}\text{Ne} \;\rightarrow\; ^{19}_{9}\text{F} \;+\; ^{0}_{1}\beta^{+}$$

b. Neon-24 is a beta emitter with a half-life of about 200 seconds. Write a balanced nuclear equation for the loss of a beta particle from this radioisotope.

$$^{24}_{10}\text{Ne} \;\rightarrow\; ^{24}_{11}\text{Na} \;+\; ^{0}_{-1}\beta$$

c. Beginning with 2 μg of neon-24, how many micrograms would remain after 200 seconds?

a.   First, calculate how many half-lives are equivalent to 200 seconds

$$\text{number of half-lives} \;=\; 200 \text{ seconds } \times \; \frac{1 \text{ half-life}}{20 \text{ seconds}} \;=\; 10 \text{ half-lives}$$

Then, use a table to determine how many micrograms of neon-24 remain after 10 half-lives:

Original amount	2.00 µg
After 1 half-life	1 µg remain
After 2 half-lives	0.5 µg remain
After 3 half-lives	0.25 µg remain
After 4 half-lives	0.125 µg remain
After 5 half-lives	0.063 µg remain
After 6 half-lives	0.031 µg remain
After 7 half-lives	0.016 µg remain
After 8 half-lives	0.0078 µg remain
After 9 half-lives	0.0039 µg remain
After 10 half-lives	**0.20 µg remain**

d. How many half –lives of neon-19 will pass in 200 seconds?

10 half-lives as calculated in part c above

2.97   *a. In chemical terms, why can exposure to nuclear radiation be harmful?*

Exposure to nuclear radiation is harmful to living tissues because of the kinetic energy that radioactive emissions impart to surrounding atoms. This transfer of energy can change the structure of water and important biochemical substances such as proteins, DNA, lipids, and others found within cells, disrupting normal biochemical functions.

b. What are some of the short-term effects of being exposed to a high dose of radiation?

The short-term effects of being exposed to a high dose of radiation include nausea to death within a few weeks depending on the dose.

c. What might be the source of a high dose of radiation?

Accidents in nuclear power plants that may cause release of high doses of radiation.

2.99   a. *Which is more easily shielded against, alpha particles, beta particles, or gamma rays?*

alpha particles

b. *Which of the types of radiation in part a is the most difficult to shield against?*

gamma particles

2.101 $^{52}_{26}$Fe, *a positron emitter with a half-life of 8.2 hours, is used for PET bone marrow scans.*

a. *Write a balanced nuclear equation for this radioactive decay.*

$$^{52}_{26}\text{Fe} \rightarrow ^{52}_{25}\text{Mn} + ^{0}_{1}\beta^+$$

b. *The detector used for this scan measures gamma rays. How is the release of positrons connected to the formation of gamma rays?*

When a positron is emitted by the $^{52}_{26}$Fe radioisotope, it collides with an electron from a nearby atom. This collision destroys both the positron and the electron, resulting in the release of two gamma rays.

c. *Assuming that this radioisotope is eliminated from the body only by radioactive decay, how much time must elapse before a patient contains just 25% of an administered dose of $^{52}_{26}$Fe ?*

16.4 hours. The half-life of the $^{52}_{26}$Fe radioisotope is 8.2 hours. After 1 half-life (8.2 hours), 50% of the administered dose remains. After 2 half-lives (16.4 hours), half of the remaining 50% will decay, bringing the quantity of the radioisotope inside the patient down to 25% of the administered dose.

d. *The product of $^{52}_{26}$Fe positron decay is a radioisotope that emits positrons. Write a balanced nuclear equation for the decay of this product.*

The product of the $^{52}_{26}$Fe decay is the $^{52}_{25}$Mn radioisotope. The decay by positron emission of this radioisotope is described by:

$$^{52}_{25}\text{Mn} \rightarrow ^{52}_{24}\text{Cr} + ^{0}_{1}\beta^+$$

2.103 *What is brachytherapy?*

Brachytherapy is a type of cancer treatment in which a radiation source is placed internally near a tumor. The radiation source delivers a high dose of radiation over time killing cancer cells in the process.

2.105 *How are fission and fusion different?*

Fission is the process by which an atom's nucleus splits to produce two smaller nuclei, a number of neutrons, and energy. Fusion is the process by which small nuclei combine, releasing a lot of energy in the process.

2.107 *Uranium – 235 can take part in other fission reaction than the one shown earlier in the chapter. Two of these nuclear reactions are shown below. Supply the missing product for each.*

$$^{235}_{92}U \ + \ ^{1}_{0}n \ \rightarrow \ ^{103}_{42}Mo \ + \ ? \ + \ 2^{1}_{0}n$$

$$^{235}_{92}U \ + \ ^{1}_{0}n \ \rightarrow \ ^{91}_{36}Kr \ + \ ? \ + \ 3^{1}_{0}n$$

Determine what the atomic number and the mass number of the missing product are by balancing the equation.

$^{131}_{50}Sn$ is the missing product for the first nuclear reaction. For the first reaction,

$$^{235}_{92}U \ + \ ^{1}_{0}n \ \rightarrow \ ^{103}_{42}Mo \ + \ ? \ + \ 2^{1}_{0}n$$

the atomic number of the missing product is calculated from $92 + 0 = 42 + ? + (2 \times 0)$. In this case, the atomic number of the missing product (?) is 50. The mass number can be calculated from $235 + 1 = 103 + ? + (2 \times 1)$ which gives 131. Using the atomic number to determine the element, the missing product is therefore:

$$^{235}_{92}U \ + \ ^{1}_{0}n \ \rightarrow \ ^{103}_{42}Mo \ + \ ^{131}_{50}Sn \ + \ 2^{1}_{0}n$$

Using the same procedure as above, the missing product for the second nuclear reaction is $^{142}_{46}Ba$:

$$^{235}_{92}U \ + \ ^{1}_{0}n \ \rightarrow \ ^{91}_{36}Kr \ + \ ^{142}_{46}Ba \ + \ 3^{1}_{0}n$$

2.109　*Iodine-131, a beta emitter with a half-life of 8.1 days, is a product of nuclear fission. After the Chernobyl nuclear accident, iodine-131 and other radioisotopes were released into the atmosphere. Almost immediately, the authorities began distributing tablets that contained the stable (non-radioactive) isotope iodine-127. What benefit would this provide the local population?*

Ingesting tablets that contain a stable isotope of iodine can help minimize the amount of the radioactive iodine taken up by biochemical systems that make use of iodine. This can help accelerate the body's ability to eliminate the radioactive iodine isotope.

2.111　*a. The United States stopped reprocessing fuel rods in the 1970s because of safety concerns. Since that time, spent fuel rods have been stored at nuclear reactor sites. What might those safety concerns have been?*

Reprocessing can yield radioactive material suitable for use in nuclear weapons. If security at a reprocessing plant is lax, or if the spent fuel rods must be transported to a reprocessing site, it might be possible for some individual or group to steal enough material to make a nuclear weapon.

b. In an energy plan released in 2006, it was proposed that reprocessing be restarted in the United States and the United States would accept spent fuel rods from other countries. What would be some benefits of this proposed plan?

Reprocessing would reduce the amount of radioactive material stored at nuclear reactor sites.

2.113　*Of the elements listed in Table 2.3, which are bulk elements and which are trace elements?*

Ca and P are bulk elements and the rest are trace elements: Fe, I, Mn, Se, and Zn.

2.115　*In NAS dietary tables the dietary reference intakes for selenium and iodine are reported in micrograms per day, rather than the milligrams per day given in Table 2.3. For these two elements, convert the values in Table 2.3 into micrograms per day.*

For selenium, the RDA given in Table 2.3 is 0.055 mg/day.

$$0.055\,\frac{\text{mg}}{\text{day}} \times \frac{10^{-3}\ \text{g}}{1\ \text{mg}} \times \frac{1\ \mu\text{g}}{10^{-6}\ \text{g}} = 55\,\frac{\mu\text{g}}{\text{day}}$$

For selenium, the UL given in Table 2.3 is 0.40 mg/day.

$$0.40\,\frac{\text{mg}}{\text{day}} \times \frac{10^{-3}\ \text{g}}{1\ \text{mg}} \times \frac{1\ \mu\text{g}}{10^{-6}\ \text{g}} = 4.0 \times 10^{2}\,\frac{\mu\text{g}}{\text{day}}$$

For iodine, the RDA given in Table 2.3 is 0.15 mg/day

$$0.15\,\frac{\text{mg}}{\text{day}} \times \frac{10^{-3}\ \text{g}}{1\ \text{mg}} \times \frac{1\ \mu\text{g}}{10^{-6}\ \text{g}} = 150\,\frac{\mu\text{g}}{\text{day}}$$

For iodine, the UL given in Table 2.3 is 1.1 mg/day

$$1.1\,\frac{\text{mg}}{\text{day}} \times \frac{10^{-3}\ \text{g}}{1\ \text{mg}} \times \frac{1\ \mu\text{g}}{10^{-6}\ \text{g}} = 1.1 \times 10^{3}\,\frac{\mu\text{g}}{\text{day}}$$

2.117   *How are electrons involved in the production of light by luciferin?*

When luciferin is acted on by a particular enzyme in the presence of oxygen gas and ATP, electrons in luciferin are raised to an excited state and, upon returning to the ground state, light is emitted.

2.121   *In late 2006, a former Russian spy was poisoned when a small amount of polonium – 210 was put in his food. He died a few weeks later.*

a.   *What is the atomic symbol of polonium?*

Po

b.   *How many protons and neutrons does an atom of polonium-210 have?*

84 protons and 126 neutrons

c. *Describe the structure of an alpha particle.*

An alpha particle has 2 protons, 2 neutrons, and a 2+ charge.

d. *Polonium-210 emits alpha particles. Write a balanced nuclear reaction for this process.*

$$^{210}_{84}\text{Po} \rightarrow \, ^{4}_{2}\alpha \, + \, ^{206}_{82}\text{Pb}$$

e. *Polonium-210 has a half-life of 138 days. Of a 2 μg sample, how much would remain after 5 half-lives?*

See table below:

$^{210}_{84}\text{Po}$ (half-life = 138 days)

Original amount	2 μg remain
After the 1st half-life	1 μg remain
After the 2nd half-life	0.5 μg remain
After the 3rd half-life	0.25 μg remain
After the 4th half-life	0.125 μg remain
After the 5th half-life	**0.06 μg remain**

f. *If alpha particles can be blocked by a sheet of paper (Figure 2.20), why is polonium – 210 so poisonous?*

The alpha particles are capable of breaking certain chemical bonds within the body which could affect important biochemical processes. Since polonium-210 was *ingested* by the victim the fact that the alpha particles do not penetrate well does not reduce their harmful effects once inside the body.

2.123 *What are the pros and cons (both long term and short term) of using radiation to treat cancer?*

The high energy associated with radiation has been shown to be effective in killing cancerous cells and mitigating the progress and symptoms of this disease. In combination with other cancer therapies like chemotherapy, it is a powerful tool for treating cancer patients. However, because radiation does not discriminate between healthy versus cancerous cells, healthy cells are killed as well, which could lead to undesirable health effects. Radiation therapy also comes with other side effects: hair loss, nausea, and loss of white blood cells. It also has some disadvantages to health care workers who administer radiation therapy because of the risks of exposure to radiation.

Chapter 3
Compounds

Solutions to Problems

3.1 *Which picture represents molecules and which one represents an ionic compound?*

Ionic compounds consist of a crystal lattice or array of alternating cations and anions, like the drawing below.

ionic compound

Molecules consist of distinct groups of atoms that are covalently bonded to each other, like the particles pictured below.

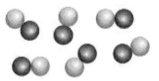

molecules

3.3 *Give the total number of protons and electrons in each ion.*

The number of protons is equal to the atomic number of the element (refer to a periodic table). Neutral atoms contain the same number of protons and electrons. Cations contain one fewer electron for each positive charge. Anions contain one more electron for each negative charge. The charge on the ion indicates the number of electrons lost or gained. For positive ions, subtract this charge value from the atomic number to calculate the number of electrons. For negative ions, add this number to the atomic number to calculate the number of electrons.

a. K^+ 19 protons $19 - 1 = 18$ electrons

b. Mg^{2+} 12 protons $12 - 2 = 10$ electrons

c. P^{3-} 15 protons $15 + 3 = 18$ electrons

3.5   *Give the total number of protons and electrons in each ion.*

a. Fe^{2+} 26 protons $26 - 2 = 24$ electrons

b. Fe^{3+} 26 protons $26 - 3 = 23$ electrons

c. Cu^+ 29 protons $29 - 1 = 28$ electrons

d. Cu^{2+} 29 protons $29 - 2 = 27$ electrons

3.7   *Give the total number of protons, neutrons, and electrons in each ion.*

a. $^{63}_{29}Cu^+$

29 protons $63 - 29 = 34$ neutrons $29 - 1 = 28$ electrons

b. $^{19}_{9}F^-$

9 protons $19 - 9 = 10$ neutrons $9 + 1 = 10$ electrons

c. $^{37}_{17}Cl^-$

17 protons $37 - 17 = 20$ neutrons $17 + 1 = 18$ electrons

3.9   *Identify the monoatomic ion that has*

Use the number of protons (the same as the atomic number) to identify the element. Use the difference between the number of protons and the number of electrons to determine the net charge on the ion.

a. 15 protons and 18 total electrons.

P^{3-} 15 protons $15 + 3 = 18$ total electrons

b. 20 protons and 18 total electrons.

Ca^{2+} 20 protons $20 - 2 = 18$ total electrons

c. 7 protons and 8 valence electrons.

N^{3-} 7 protons, 5 valence electrons $5 + 3 = 8$ valence electrons

d. 25 protons and 22 total electrons.

Mn^{3+} 25 protons $25 - 3 = 22$ total electrons

3.11   *Give the name of each ion.*

a. F^-   fluoride ion

The element is fluorine. As an anion, "ine" is replaced with "ide"

b. O^{2-}   oxide ion

The element is oxygen. As an anion, "ygen" is replaced with "ide"

c. Cl^-   chloride ion

The element is chloride. As an anion, "ine" is replaced with "ide".

d. Br^-   bromide ion

The element is bromine. As an anion, "ine" is replaced with "ide".

3.13   *Give a name for each ion.*

All of these elements are transition metal elements. When naming transition metal ions, the charge of the ion is indicated as a Roman numeral enclosed in parentheses.

a. Co^{2+} cobalt(II) ion

b. Pb^{2+} lead(II) ion

c. Cr^{3+} chromium(III) ion

d. Cu^{+} copper(I) ion

3.15   *Give the name of each ion.*

The names of polyatomic ions cannot be predicted in the same way as monatomic ions. Refer to the Table of Common Polyatomic Ions in the text.

a. $CO_3{}^{2-}$ carbonate ion

b. NO_3^- nitrate ion

c. $SO_3{}^{2-}$ sulfite ion

d. $CH_3CO_2^-$ acetate ion

3.17   *Write the formula of each ion.*

These are polyatomic ions. Formulas cannot be predicted in the same way as monatomic ions. Refer to the Table of Common Polyatomic Ions in the text.

a. *hydrogen carbonate ion* HCO_3^-

b. *nitrite ion* NO_2^-

c. *sulfate ion* $SO_4{}^{2-}$

3.19   *Draw the electron dot structure of a hydrogen cation.*

A hydrogen cation is a hydrogen atom from which the lone electron has been removed:

$$H^+$$

3.21   *How many valence electrons must each atom gain to reach an octet?*

Add as many electrons to the number of valence electrons each atom already has to determine the number required to acquire 8 valence electrons.

     a. O   6 valence electrons   8 – 6 = 2 valence electrons needed

     b. F   7 valence electrons   8 – 7 = 1 valence electron needed

     c. Se   6 valence electrons   8 – 6 = 2 valence electrons needed

     d. Br   7 valence electrons   8 – 7 = 1 valence electron needed

3.23   *Draw the electron dot structure for the ion expected to be formed from each of the atoms in Problem 3.21. Be sure to indicate the sign (positive or negative) and magnitude of the charge on the ion.*

Electron dot structures show the number of valence electrons that a given atom has. According to the octet rule, an atom will gain, lose, or share electrons to get a complete set of eight valence electrons. Each atom in Problem 3.21 needs to gain valence electrons to reach an octet. The charge is negative and the magnitude is equal to the number of valence electrons required.

    a. $:\!\ddot{\mathrm{O}}\!:^{2-}$ b. $:\!\ddot{\mathrm{F}}\!:^{1-}$ c. $:\!\ddot{\mathrm{Se}}\!:^{2-}$ d. $:\!\ddot{\mathrm{Cl}}\!:^{1-}$

3.25   *When a nitrogen atom is converted into an ion,*

a. what is the name of the ion?

nitride ion

b. how many electrons does nitrogen gain?

3 electrons. Nitrogen has 5 valence electrons and gains 3 valence electrons to reach an octet.

c. the ion ends up with what charge?

The charge on the nitride ion is 3-.

3.27   *Draw the electron dot structure of each atom and of the ion that it is expected to form.*

a. **Na·**   **Na$^+$**

Losing one electron leaves a positively charged Na atom with an octet of valence electrons.

b. **:C̈l·**   **:C̈l:$^-$**

Gaining one electron creates a chloride ion with a 1- charge and an octet of valence electrons.

c. **:Är:**   no ion formed

An argon atom already has an octet of electron and therefore does not need to gain or lose any electrons.

3.29   *Name each ionic compound.*

a. MgO   magnesium oxide

MgO is combination of magnesium ions (Mg^{2+}) and oxide ions (O^{2-}).

b. Na$_2$SO$_4$ sodium sulfate

Na_2SO_4 is a combination of sodium ions (Na^+) and sulfate ions (SO_4^{2-}). The number of times that each ion appears in the formula is not specified in the name of the compound.

c. CaF$_2$ calcium fluoride

CaF_2 is a combination of calcium ions (Ca^{2+}) and fluoride ions (F^-).

d. Na₂S   sodium sulfide

Na_2S is a combination of sodium ions (Na^+) and sulfide ions (S^{2-}).

3.31   *Name each ionic compound.*

When naming ionic compounds that contain transition metal cations, the charge of the cation is indicated in the name as a Roman numeral enclosed in parentheses.

a.   *FeCl₂*   iron(II) chloride

b.   *CoS*   cobalt(II) sulfide

c.   *CoCl₃*   cobalt(III) chloride

d.   *Al₂S₃*   aluminum sulfide (aluminum is not a transition metal)

3.33   *Write the formula of each ionic compound.*

a. calcium hydrogenphosphate $CaHPO_4$
Calcium ion is Ca^{2+} and the polyatomic ion hydrogen phosphate is HPO_4^{2-}. For the compound to be neutral, an equal number of Ca^{2+} and HPO_4^{2-} are required.

b. copper(II) bromide $CuBr_2$
Copper(II) ion is Cu^{2+} and bromide ion is Br^-. For the compound to be neutral, twice as many Br^- ions as Cu^{2+} are required.

c. copper(II) sulfate $CuSO_4$
Copper(II) ion is Cu^{2+} and sulfate ion is SO_4^{2-}. For the compound to be neutral, an equal number of Cu^{2+} and SO_4^{2-} are required.

d. sodium hydrogen sulfate $NaHSO_4$
Sodium ion is Na^+ and the polyatomic ion hydrogen sulfate is HSO_4^-. For the compound to be neutral, an equal number of Na^+ and HSO_4^- are required.

3.35   *Give the name of each ionic compound.*

      a. Li_2SO_4 (an antidepressant)

      lithium sulfate
      Li^+ is the lithium ion and SO_4^{2-} is the sulfate ion.

      b. $Ca(H_2PO_4)_2$ (used in foods as a mineral supplement)

      calcium dihydrogen phosphate
      Ca^{2+} is the calcium ion and $H_2PO_4^-$ is the dihydrogen phosphate ion.

      c. $BaCO_3$

      barium carbonate
      Ba^{2+} is the barium ion and CO_3^{2-} is the carbonate ion.

3.37   *Write the formula of the ionic compound that forms between each pair.*

      a. magnesium ions and fluoride ions MgF_2
      Twice as many (F^-) ions as magnesium ions (Mg^{2+}) are required to give a neutral compound.

      b. potassium ions and bromide ions   KBr
      An equal number of potassium ions (K^+) and bromide ions (Br^-) give a neutral compound.

      c. potassium ions and sulfide ions K_2S
      Twice as many potassium ions (K^+) as sulfide ions (S^{2-}) are required to form a neutral compound.

      d. aluminum ions and sulfide ions Al_2S_3
      To obtain a neutral compound, for every two aluminum ions (Al^{3+}), three sulfide ions (S^{2-}) are required.

3.39   *In addition to sugars, citric acid, and other ingredients, the sports drink called Powerade contains potassium phosphate and potassium dihydrogen phosphate. Write the formula of each of these ionic compounds.*

      potassium phosphate K_3PO_4

      potassium dihydrogen phosphate KH_2PO_4

3.41   *Predict the number of covalent bonds formed by each nonmetal atom.*

The number of covalent bonds that a nonmetal atom usually forms is the same as the number of electrons required by it to achieve an octet.

a. N   3 covalent bonds
A nitrogen atom (group 5A) has five valence electrons and needs 3 more for an octet.

b. Cl   1 covalent bond
A chlorine atom (group 7A) has seven valence electrons and needs 1 more for an octet.

c. P   3 covalent bonds
A phosphorus atom (group 5A) has five valence electrons and needs 3 more for an octet.

3.43   *Draw the electron dot structure of each molecule.*

In the molecular drawings given, replace each single bond with two dots (:)and each double bond with four dots (::).

a.

$$\begin{array}{ccc} & H & \overset{..}{} & H \\ & | & & | \\ H- & C-O-C & -H \\ & | & \overset{..}{} & | \\ & H & & H \end{array}$$

$$\begin{array}{ccc} & H & H \\ & \overset{..}{} & \overset{..}{} \\ H\!:\!C\!:\!O\!:\!C\!:\!H \\ & \overset{..}{} & \overset{..}{} \\ & H & H \end{array}$$

b.

$$\begin{array}{cccc} H & H & H & :O \\ | & | & | & || \\ H-C & -C & -C-C-H \\ | & | & & \\ H & H-N: & H \\ & | & \\ & H & \end{array}$$

$$\begin{array}{cccc} & & & \overset{..}{O}: \\ H & H & H & :: \\ H\!:\!C & : & C\!:\!C\!:\!C\!:\!H \\ & & & \\ H & H\!:\!N: & H \\ & : & \\ & H & \end{array}$$

53

3.45 *On the structure shown in Problem 3.44a, point out a pair of bonding electrons.*

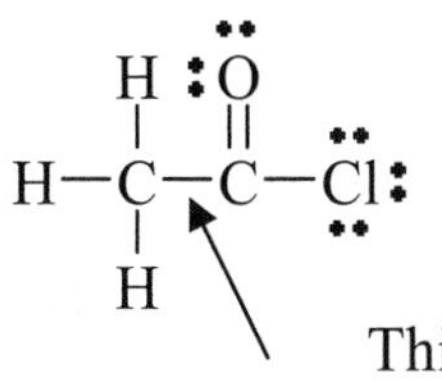

This is an example of a bonding pair. Other examples include the C-Cl bond, all C-H bonds, and each of the 2 C-O bonds.

3.47 *Draw the line-bond structure of pyruvic acid, a compound formed during the breakdown of sugars by the body.*

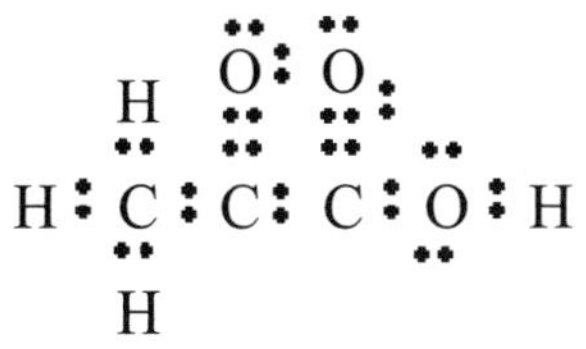

Pyruvic acid

Substitute a line for each pair of dots that represents a bond.

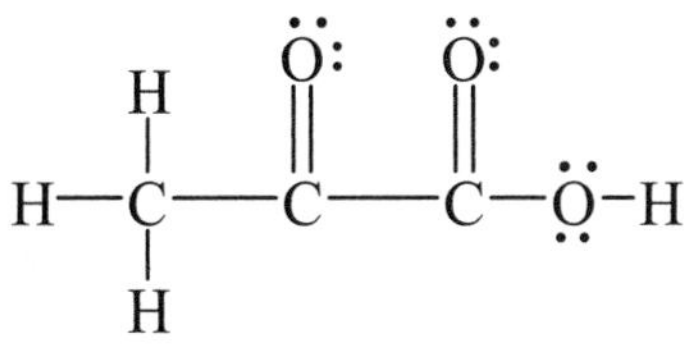

3.49 *Name each binary molecule.*

When naming binary molecules, the relative number of each atom is specified using a prefix. (mono = 1, di = 2, tri = 3, tetra =4, penta = 5, and hexa = 6) Note that the mono prefix is not used on the first element written when only one atom is present.

a. NCl₃   nitrogen trichloride

b. PCl₃   phosphorus trichloride

c. PCl₅   phosphorus pentachloride

3.51   *Phosphine (PH₃) is a poisonous gas that has the odor of decaying fish. Give another name for this binary molecule.*

phosphorus trihydride

3.53   *Draw the electron dot structure of the molecule formed when sufficient H atoms are added to give each atom an octet of valence electrons.*

Atoms will form as many covalent bonds as are required to acquire eight valence electrons.

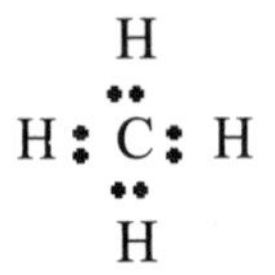

a. C

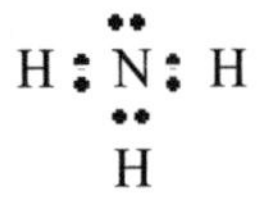

$$H \!:\! \overset{\displaystyle H}{\underset{\displaystyle H}{\overset{\bullet\bullet}{\underset{\bullet\bullet}{C}}}} \!:\! H$$

b. N

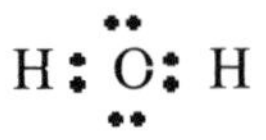

$$H \!:\! \overset{\bullet\bullet}{\underset{\displaystyle H}{\underset{\bullet\bullet}{N}}} \!:\! H$$

c. O

$$H \!:\! \overset{\bullet\bullet}{\underset{\bullet\bullet}{O}} \!:\! H$$

3.55   *Indicate whether each is an ionic compound or a binary molecule.*

Ionic compounds contain at least one metallic element. A binary molecule contains only nonmetallic elements.

a. $BaCl_2$	ionic		d. HgO	ionic	
b. OCl_2	binary molecule		e. N_2O_3	binary molecule	
c. CS_2	binary molecule		f. Cu_2O	ionic	

3.57   *Name each of the ionic compounds or binary molecules in Problem 3.55.*

a. $BaCl_2$	barium chloride		d. HgO	mercury(II) oxide
b. OCl_2	oxygen dichloride		e. N_2O_3	dinitrogen trioxide
c. CS_2	carbon disulfide		f. Cu_2O	copper(I) oxide

3.59   *Are all diatomic molecules compounds? Explain*

No. A diatomic molecule consists of two atoms bonded together. If the atoms are of different elements (such as CO), then the diatomic molecule is a compound. If the atoms are the same (such as N_2), then the diatomic molecule is an element.

3.61   *a. What is the formula weight of ammonium hydroxide?*

The formula weight of a compound can be calculated by adding the atomic weights of all the atoms of each element in the formula. Ammonium hydroxide has the formula NH_4OH.

Formula weight of NH_4OH =

(1 x 14.01 amu) + (5 x 1.01 amu) + (1 x 16.00 amu) = 35.06 amu

b. What is the mass of 0.950 mol of ammonium hydroxide?

Use the molar mass of NH_4OH (35.06 g/mol) as a conversion factor to calculate the mass of 0.950 mol of NH_4OH.

$$0.950 \; \cancel{mol \; NH_4OH} \; \times \; \frac{35.06 \text{ g } NH_4OH}{1 \; \cancel{mol \; NH_4OH}} \; = \; 33.3 \text{ g } NH_4OH$$

c. How many moles of ammonium hydroxide are present in 0.475 g?

To calculate the moles of NH4OH present in 0.475 g of NH4OH, use the molar mass as a conversion factor as shown below.

$$0.475 \text{ g NH}_4\text{OH} \times \frac{1 \text{ mol NH}_4\text{OH}}{35.06 \text{ g NH}_4\text{OH}} = 0.135 \text{ mol NH}_4\text{OH}$$

3.63   *a. What is the formula weight of sodium oxide?*

Sodium oxide has the formula Na_2O.

Formula weight of Na_2O = (2 x 23.00 amu) + (1 x 16.00 amu) = 62.00 amu

b. How many oxide ions are present in 0.25 mol of sodium oxide?

Use the ratio given by the chemical formula of the compound to determine the amount of a component in a compound. In every 1 mole of Na_2O, there is 1 mole of O^{2-} ions.

$$0.25 \text{ mol Na}_2\text{O} \times \frac{1 \text{ mol O}^{2-}}{1 \text{ mol Na}_2\text{O}} \times \frac{6.02 \times 10^{23} \text{ O}^{2-} \text{ ions}}{1 \text{ mol O}^{2-}} = 1.5 \times 10^{23} \text{ O}^{2-} \text{ ions}$$

c. How many sodium ions are present in 0.25 mol of sodium oxide?

In every 1 mole of Na_2O, there are 2 moles of Na^+ ions.

$$0.25 \text{ mol Na}_2\text{O} \times \frac{2 \text{ mol Na}^+}{1 \text{ mol Na}_2\text{O}} \times \frac{6.02 \times 10^{23} \text{ Na}^+ \text{ ions}}{1 \text{ mol Na}^+} = 3.0 \times 10^{23} \text{ Na}^+ \text{ ions}$$

d. How many oxide ions are present in 2.30 g of sodium oxide?

First, convert the mass of the sample to moles of the sample. Then, proceed as above.

$$2.30 \text{ g Na}_2\text{O} \times \frac{1 \text{ mol Na}_2\text{O}}{62.00 \text{ g Na}_2\text{O}} \times \frac{1 \text{ mol O}^{2-}}{1 \text{ mol Na}_2\text{O}} \times \frac{6.02 \times 10^{23} \text{ O}^{2-} \text{ ions}}{1 \text{ mol O}^{2-}} = 2.23 \times 10^{22} \text{ O}^{2-} \text{ ior}$$

e. How many sodium ions are present in 2.30 g of sodium oxide?

$$2.30 \ \cancel{g \ Na_2O} \ \times \ \frac{1 \ \cancel{mol \ Na_2O}}{62.00 \ \cancel{g \ Na_2O}} \ \times \ \frac{2 \ \cancel{mol \ Na^+}}{1 \ \cancel{mol \ Na_2O}} \ \times \ \frac{6.02 \times 10^{23} \ Na^+ \ ions}{1 \ \cancel{mol \ Na^+}} = \ 4.47 \times 10^{22} \ Na^+ \ ions$$

3.65 *The food additive potassium sorbate, $K(C_6H_7O_2)$, is a mold and yeast inhibitor.*

a. What is the charge on the sorbate ion?

1−. The charge on the potassium ion is 1+. Because the formula indicates a one-to-one ratio between potassium and sorbate, the charge on the sorbate ion must be 1−.

b. How many C atoms are present in 0.0150 g of potassium sorbate?

First, calculate the formula weight in grams of the potassium sorbate compound. Then use this to convert the mass of the sample to moles of the sample. Use the ratio given by the formula and Avogadro's number to determine the number of C atoms as shown below:

$$0.0150 \ \cancel{g \ KC_6H_7O_2} \ \times \ \frac{1 \ \cancel{mol \ KC_6H_7O_2}}{150.22 \ \cancel{g \ KC_6H_7O_2}} \ \times \ \frac{6 \ mol \ C}{1 \ \cancel{mol \ KC_6H_7O_2}}$$

$$\times \ \frac{6.02 \times 10^{23} \ C \ atoms}{\cancel{1 \ mol \ C}} = \ 3.61 \times 10^{20} \ C \ atoms$$

3.67 *a. What is the molecular weight of CCl_4?*

molecular weight of CCl_4 = (1 x 12.01 amu) + (4 x 35.45 amu) = 153.81 amu

b. What is the mass of 61.3 mol of CCl_4?

$$61.3 \ \cancel{mol \ CCl_4} \ \times \ \frac{153.81 \ g \ CCl_4}{1 \ \cancel{mol \ CCl_4}} \ = \ 9.43 \times 10^3 \ g \ CCl_4$$

c. How many moles of CCl_4 are present in 0.465 g of CCl_4?

$$0.465 \; \text{g } CCl_4 \quad \text{x} \quad \frac{1 \text{ mol } CCl_4}{153.81 \; \text{g } CCl_4} \quad = \quad 3.02 \text{ x } 10^{-3} \text{ mol } CCl_4$$

d. How many molecules of CCl_4 are present in 5.50 x 10^{-3} g of CCl_4?

$$5.50 \text{ x } 10^{-3} \; \text{g } CCl_4 \quad \text{x} \quad \frac{1 \text{ mol } CCl_4}{153.81 \; \text{g } CCl_4} \quad \text{x} \quad \frac{6.02 \text{ x } 10^{23} \; CCl_4 \text{ molecules}}{1 \text{ mol } CCl_4} = \quad 2.15 \text{ x } 10^{19} \; CCl_4 \text{ molecules}$$

3.69   *One tablet of a particular analgesic contains 250 mg acetaminophen ($C_8H_9NO_2$). How many acetaminophen molecules are contained in the tablet?*

First convert milligrams into grams, then proceed as in previous problems.

$$250 \; \text{mg} \quad \text{x} \quad \frac{1 \text{ x } 10^{-3} \; \text{g}}{1 \; \text{mg}} \quad \text{x} \quad \frac{1 \text{ mol}}{151.17 \text{ g}} \quad \text{x} \quad \frac{6.02 \text{ x } 10^{23} \text{ molecules}}{1 \text{ mol}} \quad = \quad 1.0 \text{ x } 10^{21} \text{ molecules}$$

3.71   *Isovaleric acid ($C_5H_{10}O_2$) is the molecule responsible for foot odor. To smell this compound, it must be present in the air at a minimum of 250 parts per billion. This means that a 0.50 L volume of air would hold 0.00013 g of isovaleric acid.*

a. To how many moles of isovaleric acid does this correspond?

$$0.00013 \; \text{g } C_5H_{10}O_2 \quad \text{x} \quad \frac{1 \text{ mol } C_5H_{10}O_2}{102.13 \; \text{g } C_5H_{10}O_2} \quad = \quad 1.3 \text{ x } 10^{-6} \text{ mol } C_5H_{10}O_2$$

b. To how many molecules of isovaleric acid does this correspond?

$$1.3 \text{ x } 10^{-6} \text{ mol } C_5H_{10}O_2 \quad \text{x} \quad \frac{6.02 \text{ x } 10^{23} \text{ molecules } C_5H_{10}O_2}{1 \text{ mol } C_5H_{10}O_2} \quad = \quad 7.8 \text{ x } 10^{17} \text{ molecules}$$

3.73   *Explain how ionophores act as antibiotics.*

Ionophores are molecules that transport ions across cell membranes. Ionophores can be used to kill bacteria (as antibiotics) by forcing the passage of ions across the cell membrane of the bacteria and disrupting the balance of ions which hampers the normal biological functions of the bacterial cell.

3.75   *Sometimes a filling can make a tooth sensitive to temperature changes. Given that amalgam is made from metals and that composites are made from nonmetals, which type of filling would you expect is most likely to cause thermal sensitivity?*

The amalgam would likely result in increased thermal sensitivity because it conducts heat more readily than composites.

3.77   *Viagra, one of the drugs used to treat erectile dysfunction, works by enhancing the effects of nitric oxide produced by the body. Men who take nitroglycerin for angina are advised against taking Viagra. Why?*

Nitroglycerin reduces constriction of arteries (angina) by producing nitric oxide which causes blood vessels to dilate. Viagra enhances the effects of nitric oxide. Taking Viagra with nitroglycerin may result in a precipitous, and possibly life-threatening, drop in blood pressure when blood vessel dilation is prolonged or severe.

3.81   *Some ionic compounds can form hydrates, in which water molecules are incorporated into their crystal structures. One example is iron(II) nitrate hexahydrate, $Fe(NO_3)_2 \cdot 6H_2O$. The dot in the formula indicates that there are 6 water molecules associated with each formula unit of iron(II) nitrate. Because water molecules are neutral, they contribute no charges to the compound.*

a. What is the charge on the cation in this compound?

2+.   The charge on each of the nitrate ions is 1-. Two nitrate ions to make a neutral compound means that the charge on the iron must be 2+.

b. Give an alternate name for the cation.

ferrous ion

c. What is the charge on the anion in this compound?

1-.   The anion is the polyatomic ion NO_3^-.

d. Does the compound contain any ionic bonds? Explain.

Yes. The bonds between 1 Fe^{2+} ion and 2 NO_3^- ions are ionic.

e. Does the compound contain any covalent bonds? Explain.

Yes. The bonds between 1 N and 3 O atoms in NO_3^- and the bonds between 2 H atoms and 1 O atom in each H_2O molecule are all covalent.

f. Give an alternate name for water, using the rules for naming binary compounds.

dihydrogen monoxide

g. What is the formula weight of iron(II) nitrate hexahydrate?

Formula weight of $Fe(NO_3)_2 \cdot 6H_2O$ =

(1 x 55.85 amu) + (2 x 14.01 amu) + (12 x 16.00 amu)

$+$ (12 x 1.01 amu) = 287.99 amu

h. How many moles is 4.93 g of this compound?

$$4.93 \ \cancel{g} \ \times \ \frac{1 \ mol}{287.99 \ \cancel{g}} \ = \ 1.71 \times 10^{-2} \ mol$$

i. How many grams is 0.639 mol of this compound?

$$0.639 \ \cancel{mol} \ \times \ \frac{287.99 \ g}{1 \ \cancel{mol}} \ = \ 184 \ g$$

j. How many water molecules are in 8.58×10^{-6} mol of this compound?

$$8.58 \times 10^{-6} \ \cancel{mol \ Fe(NO_3)_2 \cdot 6H_2O} \ \times \ \frac{6 \ \cancel{mol \ H_2O}}{1 \ \cancel{mol \ Fe(NO_3)_2 \cdot 6H_2O}}$$

$$\times \ \frac{6.02 \times 10^{23} \ molecules \ H_2O}{1 \ \cancel{mol \ H_2O}} \ = \ 3.10 \times 10^{19} \ molecules \ H_2O$$

k. How many iron(II) ions are in 43.8 g of this compound?

$$43.8 \; \cancel{g} \; \cancel{Fe(NO_3)_2 \cdot 6H_2O} \; \times \; \frac{1 \; \cancel{mol \; Fe(NO_3)_2 \cdot 6H_2O}}{287.99 \; \cancel{g \; Fe(NO_3)_2 \cdot 6H_2O}}$$

$$\times \; \frac{1 \; \cancel{mol \; Fe^{2+}}}{1 \; \cancel{mol \; Fe(NO_3)_2 \cdot 6H_2O}} \; \times \; \frac{6.02 \times 10^{23} \; Fe^{2+} \; ions}{1 \; \cancel{mol \; Fe^{2+}}} \; = \; 9.16 \times 10^{22} \; Fe^{2+} \; ions$$

Chapter 4
An Introduction to Organic Compounds

Solutions to Problems

4.1   *Identify the term that best describes the two molecules below: constitutional isomers, geometric isomers, different conformations of the same molecule, two unrelated molecules.*

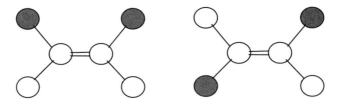

Geometric isomers.

The two molecules differ only by the configuration around the C=C. The left molecule has two identical groups on the same side of the double bond. The right molecule has two identical groups on opposite sides of the double bond.

4.3   *Indicate the number of covalent bonds that each nonmetal atom is expected to form.*

       a. C   4 bonds. Carbon has 4 valence electrons and therefore needs to share 4 more electrons to complete its octet number of electrons.

       b. O   2 bonds. Oxygen has 6 valence electrons and therefore needs to share 2 more electrons to complete its octet number of electrons.

       c. P   3 bonds. Phosphorus has 5 valence electrons and therefore needs to share 3 more electrons to complete its octet number of electrons.

       d. Br   1 bond. Bromine has 7 valence electrons and therefore needs to share only 1 more electron to complete its octet number of electrons.

4.5   *Draw the structural formula of a molecule that contains*
a. one oxygen atom and two hydrogen atoms.
b. one hydrogen atom and one iodine atom.
c. one nitrogen atom and three hydrogen atoms.

Use the procedure for drawing line-bond structures given in the text, keeping in mind that a hydrogen atom should have a pair of electrons and all other atoms in the molecule should have an octet of electrons.

a. one oxygen atom and two hydrogen atoms

$$H-\overset{\displaystyle ..}{\underset{\displaystyle ..}{O}}-H$$

b. one hydrogen atom and one iodine atom

$$H-\overset{\displaystyle ..}{\underset{\displaystyle ..}{I}}:$$

c. one nitrogen atom and three hydrogen atoms

$$H-\overset{\displaystyle ..}{\underset{\displaystyle |}{N}}-H$$
$$H$$

4.7 *Draw the line-bond structure of each molecule.*

Use the procedure for drawing line-bond structures given in the text, keeping in mind that each atom in the diatomic molecule should have an octet of electrons.

a. F₂

$$:\overset{\displaystyle ..}{\underset{\displaystyle ..}{F}}-\overset{\displaystyle ..}{\underset{\displaystyle ..}{F}}:$$

b. O₂

$$:\overset{\displaystyle ..}{O}=\overset{\displaystyle ..}{O}:$$

4.9 *Draw the line-bond structure of each molecule.*

a. CH₂S

$$\overset{\displaystyle ..}{S}:$$
$$\|$$
$$H-C-H$$

b. NF₃

$$:\overset{\displaystyle ..}{\underset{\displaystyle ..}{F}}-\overset{\displaystyle}{\underset{\displaystyle |}{N}}-\overset{\displaystyle ..}{\underset{\displaystyle ..}{F}}:$$
$$:\overset{\displaystyle}{\underset{\displaystyle ..}{F}}:$$

4.11   *Draw each molecule.*

Using line-bond structures:

a. C_2H_6

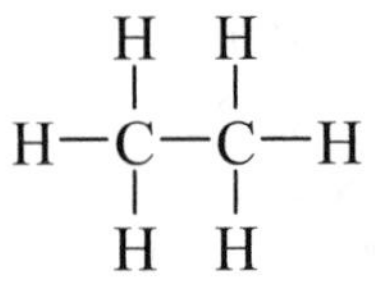

$$\begin{array}{ccc} H & & H \\ | & & | \\ H-C-&C&-H \\ | & & | \\ H & & H \end{array}$$

b. C_2H_4

H—C=C—H
   | |
   H   H

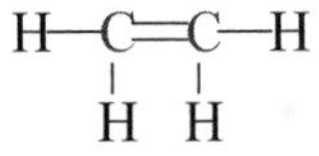

c. C_2H_2

H—C≡C—H

4.13   *Draw each polyatomic ion. Each atom, except for hydrogen, should have an octet of valence electrons.*

a. OH^-

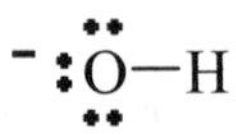

$^-$:Ö—H

b. NH_4^+

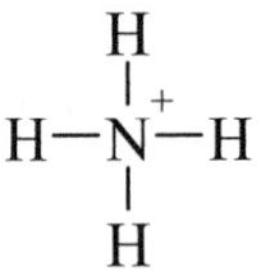

$$\begin{array}{c} H \\ | \\ H-\overset{+}{N}-H \\ | \\ H \end{array}$$

c. CN^-

$^-$:C≡N:

4.15   *Draw each of the following. Each atom should have an octet of valence electrons.*

a. SO_3

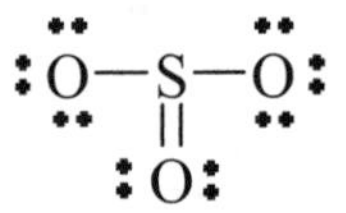

:Ö—S—Ö:
     ‖
    :Ö:

b. SO_3^{2-}

:Ö—S—Ö:$^{2-}$
     |
    :Ö:

65

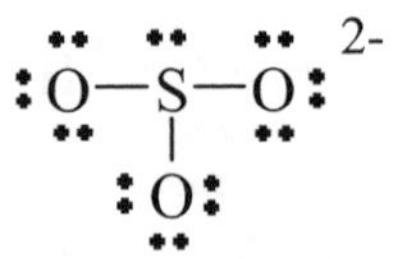

4.17 *Draw two different molecules that have the formula C₂H₆O.*

$$H-\overset{\overset{\displaystyle H}{|}}{\underset{\underset{\displaystyle H}{|}}{C}}-\overset{\bullet\bullet}{\underset{\bullet\bullet}{O}}-\overset{\overset{\displaystyle H}{|}}{\underset{\underset{\displaystyle H}{|}}{C}}-H \qquad H-\overset{\overset{\displaystyle H}{|}}{\underset{\underset{\displaystyle H}{|}}{C}}-\overset{\overset{\displaystyle H}{|}}{\underset{\underset{\displaystyle H}{|}}{C}}-\overset{\bullet\bullet}{\underset{\bullet\bullet}{O}}-H$$

4.19 *Write a condensed structural formula for each molecule.*

a.
$$H-\overset{\overset{\displaystyle H}{|}}{\underset{\underset{\displaystyle H}{|}}{C}}-\overset{\overset{\displaystyle H}{|}}{\underset{\underset{\displaystyle H}{|}}{C}}-\overset{\overset{\displaystyle H}{|}}{\underset{\underset{\displaystyle H}{|}}{C}}-\overset{\overset{\displaystyle H}{|}}{\underset{\underset{\displaystyle H}{|}}{C}}-H$$

$CH_3CH_2CH_2CH_3$

b.
$$H-\overset{\overset{\displaystyle H}{|}}{\underset{\underset{\displaystyle H}{|}}{C}}-\overset{\overset{\displaystyle H}{|}}{\underset{\underset{\displaystyle H}{|}}{C}}-\overset{\bullet\bullet}{\underset{\underset{\displaystyle H}{|}}{N}}-H$$

$CH_3CH_2NH_2$

4.21 *Draw a skeletal structure for each molecule in Problem 4.19.*

a.
$$H-\overset{\overset{\displaystyle H}{|}}{\underset{\underset{\displaystyle H}{|}}{C}}-\overset{\overset{\displaystyle H}{|}}{\underset{\underset{\displaystyle H}{|}}{C}}-\overset{\overset{\displaystyle H}{|}}{\underset{\underset{\displaystyle H}{|}}{C}}-\overset{\overset{\displaystyle H}{|}}{\underset{\underset{\displaystyle H}{|}}{C}}-H$$

b.
$$H-\overset{\overset{\displaystyle H}{|}}{\underset{\underset{\displaystyle H}{|}}{C}}-\overset{\overset{\displaystyle H}{|}}{\underset{\underset{\displaystyle H}{|}}{C}}-\overset{\bullet\bullet}{\underset{\underset{\displaystyle H}{|}}{N}}-H$$

4.23 *Which atom in each pair is more electronegative?*

a. N and O   O. The electronegativity of O is 3.5 and N is 3.0.

b. O and S   O. The electronegativity of O is 3.5 and S is 2.5.

c. Na and Cl   Cl. The electronegativity of Cl is 3.0 and Na is 0.9.

4.25   *Calculate the electronegativity difference between each pair of atoms.*

        a. N and O $3.5 - 3.0 = 0.5$

        b. C and P $2.5 - 2.1 = 0.4$

        c. H and O $3.5 - 2.1 = 1.4$

        d. Li and Cl $3.0 - 1.0 = 2.0$

4.27   *For each pair of atoms in Problem 4.25, identify the bond that would form between them as nonpolar covalent, polar covalent, or ionic.*

Use the following ranges of electronegativity differences to identify the type of bond:

Less than 0.5	nonpolar covalent
Between 0.5 – 1.9	polar covalent
Greater than 1.9	ionic

        a. N and O $3.5 - 3.0 = 0.5$ polar covalent

        b. C and P $2.5 - 2.1 = 0.4$ nonpolar covalent

        c. H and O $3.5 - 2.1 = 1.4$ polar covalent

        d. Li and Cl $3.0 - 1.0 = 2.0$ ionic

4.29   *Identify each bond as being polar covalent or nonpolar covalent.*

Use the following ranges of electronegativity differences to identify the type of bond:

Less than 0.5	nonpolar covalent
Between 0.5 – 1.9	polar covalent
Greater than 1.9	ionic

        a. C and H $2.5 - 2.1 = 0.4$ nonpolar covalent

        b. C and O $3.5 - 2.5 = 1.0$ polar covalent

        c. F and F $4.0 - 4.0 = 0.0$ nonpolar covalent

4.31   *Label any polar covalent bond(s) in each molecule or ion.*

Use the following ranges of electronegativity differences to identify the type of bond:

Less than 0.5   nonpolar covalent
Between 0.5 – 1.9   polar covalent
Greater than 1.9   ionic

a. None. The electronegativity difference between H and S is 0.4.

b. Each C-O bond is polar covalent because the electronegativity difference between the atoms is 1.0.

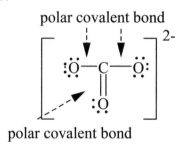

c. None. The electronegativity difference between N and Cl is 0.

4.33   *Specify the shape around each specified atom in Problem 4.31.*

a. The S atom in H_2S
b. The C atom in $CO_3{}^{2-}$
c. The N atom in NCl_3

a. Bent. The S atom has *four groups* of electrons around it: two single bonds to H atoms and two pairs of nonbonding electrons.

b. Trigonal planar. The C atom has *three groups* of electrons around it: 2 single bonds and 1 double bond to three O atoms.

c. Pyramidal. The N atom has *four groups* of electrons around it: 3 covalent bonds to Cl atoms and a nonbonding pair of electrons.

4.35   Identify the shape around each specified atom in Problem 4.9.

   a. The C atom in CH₂S.

   Trigonal planar. The C atom has *three groups* of electrons around it: two single
   bonds to H atoms and a double bond to a S atom.

   b. The N atom in NF₃.

   Pyramidal. The N atom has *four groups* of electrons around it: 3 single bonds to
   3 F atoms and a nonbonding pair.

4.37   *Which of the molecules are polar?*

a.
```
      H   ··
      |  ··
  H—C—F:
      |  ··
      H
```
b.
```
      H
      |
  ··  |  ··
 :F—C—F:
  ··  |  ··
      H
```
c.
```
        ··
       :F:
        |
  ··    |    ··
 :F—C—F:
  ··    |    ··
       :F:
        ··
```

   a and b.

   To be polar the molecule must meet two basic criteria: it must have polar bonds and
   these bonds cannot be equally distributed in the molecule (the molecule must be
   unsymmetrical). Since all of the molecules have polar bonds, the only criterion to
   check is bond distribution. The shape around the C atom in each molecule is
   tetrahedral. Only in c do the polar bonds cancel one another and make the molecule
   nonpolar.

4.39   Which, if any, of the molecules in Problem 4.9 are polar?

   a. CH₂S

   Not polar. This molecule has no polar covalent bonds.

   b. NF₃

   Polar. The N-F bonds are polar covalent and the shape of the molecule leads to
   an unsymmetrical distribution of electrons.

4.41   *True or false? A bent molecular shape always has a bond angle near 110°.*

False. Depending on the number of groups of electrons around an atom, a bent molecular shape can have a bond angle near 110° or near 120°.

4.43   *Do hydrogen bonds form between formaldehyde molecules (Table 4.2)?*

No. In order to have hydrogen bonding, there must be at least one hydrogen atom attached to a nitrogen, oxygen, or fluorine atom. In examining the formaldehyde structure, $H_2C=O$, we see that the hydrogen atoms are not attached to the oxygen, so hydrogen bonds do not form between formaldehyde molecules.

4.45   *Which pairs of molecules can form a hydrogen bond with one another?*

   a. CH_3CH_3 and CH_3CH_3

No. There are no hydrogen atoms bonded to an oxygen, nitrogen, or fluorine atom in either of the molecules. Therefore, no hydrogen bonds can form.

   b. $CH_3-\overset{\displaystyle O}{\overset{\displaystyle \|}{C}}-H$ and $CH_3-\overset{\displaystyle O}{\overset{\displaystyle \|}{C}}-H$

No. There are no hydrogen atoms attached to N, O, or F.

   c. CH_3CH_2OH and CH_3CH_2OH

Yes. The two molecules can form a hydrogen bond between them because there is a hydrogen atom in each molecule attached to an oxygen atom. The hydrogen atom from one molecule can form a hydrogen bond with the oxygen atom from the other molecule.

   d. $CH_3-\overset{\displaystyle O}{\overset{\displaystyle \|}{C}}-OH$ and $CH_3-\overset{\displaystyle O}{\overset{\displaystyle \|}{C}}-OH$

Yes. The two molecules can form a hydrogen bond between them because there is a hydrogen atom in each molecule attached to an oxygen atom. The hydrogen atom from one molecule can form a hydrogen bond with the oxygen atom from the other molecule.

4.47   *Which of the molecules in Problem 4.45 can form a hydrogen bond with a water molecule?*

b, c, and d.

A hydrogen bond is the interaction of a nitrogen, oxygen, or fluorine atom with a hydrogen atom that is covalently bonded to different nitrogen, oxygen, or fluorine atom. This gives two criteria for hydrogen bonding to occur; first, at least one of the molecules must have a hydrogen atom that is attached to a nitrogen, oxygen, or fluorine atom, second, the other molecule must have a nitrogen, oxygen, or fluorine atom in its structure. Water meets either criterion, the molecule in part b meets the second criterion, the molecules in parts c and d meet either criterion.

4.49   *A protein contains the following groups. Which can form salt bridges with one another?*

a. $-CH_2OH$ and $HOCH_2-$

b. $-CH_2\overset{\displaystyle O}{\overset{\displaystyle \|}{C}}-O^-$ and $\overset{+}{NH_3}CH_2CH_2CH_2-$

c. $-CH_2\overset{\displaystyle O}{\overset{\displaystyle \|}{C}}-OH$ and $\overset{+}{NH_3}CH_2CH_2CH_2-$

d. $-\underset{\displaystyle CH_3}{\overset{\displaystyle}{CH}}CH_3$ and CH_3-

b. only

A salt bridge is another name for an ionic bond. The term is used to describe ionic bonds that form between oppositely charged groups in protein molecules. Therefore, the basis for identifying the parts of the protein that can form a salt bridge is to identify parts of the molecule with a charge. In this case, groups shown in b are the only ones that both have opposite charges ($-O^-$ in one and the $-NH_3^+$ in the other).

4.51   *Which share the stronger London force interactions, two $CH_3CH_2CH_2CH_2CH_3$ molecules or two $CH_3CH_2CH_2CH_2CH_2CH_2CH_2CH_3$ molecules?*

Two $CH_3CH_2CH_2CH_2CH_2CH_2CH_2CH_3$ molecules.

The larger the molecule, the more electrons there are that can participate in a London force interaction. Therefore, the bigger molecules will experience stronger London force interactions.

4.53   *In ancient Greece, Socrates and others were executed by being forced to drink hemlock. The major poisonous ingredient in hemlock is an organic molecule called coniine.*

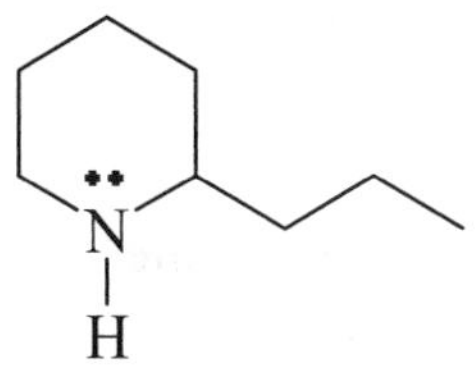

Coniine

Show two ways that a coniine molecule can form a hydrogen bond to a water molecule.

The coniine molecule contains an N-H bond that is capable of hydrogen-bonding with a water molecule in two ways as shown below:

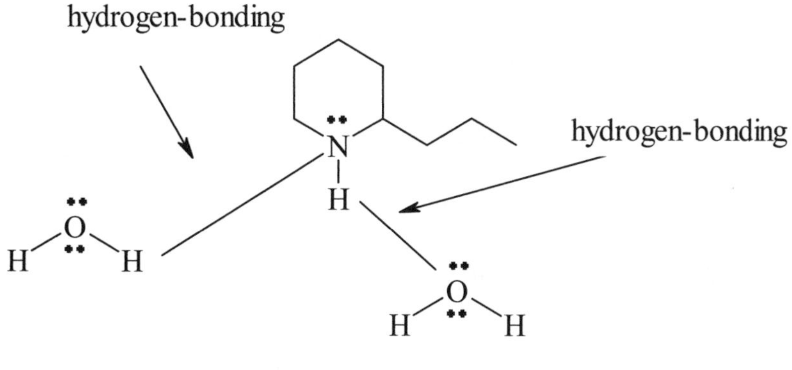

4.55   *Which has the higher boiling point, $CH_3CH_2CH_2CH_2CH_3$ or $CH_3CH(CH_3)CH_2CH_3$? Explain.*

$CH_3CH_2CH_2CH_2CH_3$.

The stronger the interactions are between the molecules in a liquid, the higher the boiling point will be. Both molecules have the formula C_5H_{12}, but $CH_3CH_2CH_2CH_2CH_3$ is unbranched and has a greater surface area. This leads to stronger London force attractions between molecules.

4.57   *Arrange the molecules in order, from the highest boiling point to lowest boiling point: decane, propane, butane.*

decane > butane > propane

Note that these are all "straight chained" molecules so that branching need not be considered. In this, the longer the hydrocarbon chain, the higher the boiling point will be. Remember that boiling points increase as the London force attraction increases. Decane (10 carbon atoms) is the longest and has the strongest London force interactions, followed by butane (4 carbon atoms) and propane (3 carbon atoms).

4.59   *Draw a line-bond structure of each alkane.*

Connect every pair of atoms in the formula with a single line to represent the covalent bond between them. Note how the (CH_3) group is written as a side branch.

a. $CH_3CH_2C(CH_3)_3$

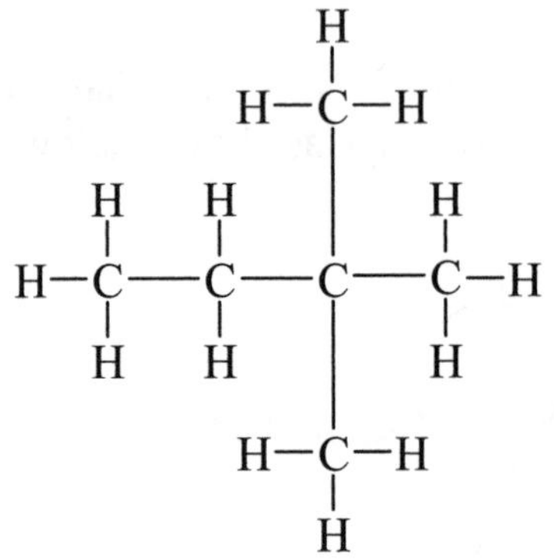

b. $CH_3CH_2CH(CH_3)CH(CH_3)CH_2CH_3$

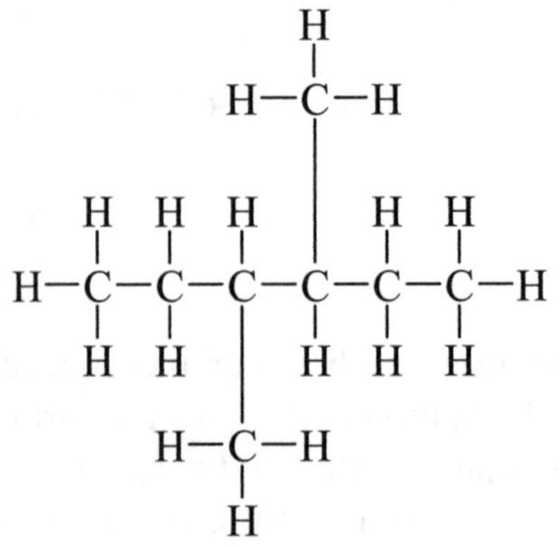

4.61   *Draw a skeletal structure for each of the molecules in Problem 4.59.*

The skeletal structure is drawn by representing all covalent bonds by lines and leaving out all of the carbon and hydrogen atoms.

a.   b.

4.63   *Draw a line bond structure for each molecule.*

a.

b.

4.65   *Write a condensed structural formula for each molecule in Problem 4.63.*

a. $CH_3CH_2CH(CH_3)_2$

b.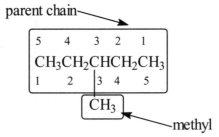

CH₃CH₂CH(CH₂CH₃)CH(CH₃)CH(CH₃)₂

$CH_3CH_2CH(CH_2CH_3)CH(CH_3)CH(CH_3)_2$

4.67   *Find and name the parent chain for each molecule, then give the complete IUPAC name for each.*

  a.   *b.*   *c.*

$$CH_2CH_2CH_3$$

$$CH_3CH_2CHCH_2CH_3 \qquad CH_3CHCH_2CH_3 \qquad CH_3CCH_2CH_3$$

$$\quad\quad\quad |\qquad\qquad\qquad\qquad\quad |\qquad\qquad\qquad\qquad |$$

$$\quad\quad CH_3 \qquad\qquad\quad CH_2CH_2CH_3 \qquad\quad CH_2CH_2CH_3$$

First, identify the parent chain by finding the longest chain of carbon. Remember that the longest chain is not necessarily the one that goes straight across. Assign the appropriate name using the numbering prefix and the ending "ane". Next, identify and name the substituents. Finally, to complete the IUPAC naming process, number the carbons of the parent chain starting from the end nearer the first substituent.

a. 3-methylpentane.
For this molecule, the longest chain has five carbon atoms, making its parent

parent chain⟍

5	4	3	2	1

CH₃CH₂CHCH₂CH₃

1	2	3	4	5

CH₃

⟵methyl

chain *pentane*. The only substituent is a methyl group and the numbering gives the same number (3) for the methyl position whether you count left to right or right to left.

b. 3-methylhexane.

75

Note how the parent chain turns and is not straight across. The longest chain

parent chain

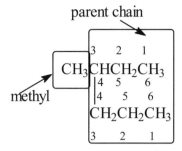

methyl

has six carbon atoms making the parent chain *hexane*. This molecule has only one substituent, the methyl group. In this case numbering top to bottom gives the methyl group a position of 3 but numbering from bottom to top gives a position of 4. The number 3 is the assigned position.

c. 4-ethyl-4-methylheptane.
Note how the parent chain turns and is not straight across. The longest

parent chain

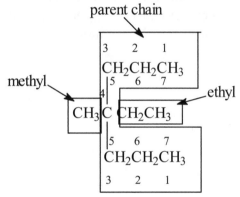

methyl

ethyl

continuous chain is seven carbon atoms long making the parent chain *heptane*. This molecule has two substituents, one methyl group and one ethyl group, given in alphabetical order in the name. The carbon they are attached to gets the same assigned number from both directions.

4.69   *Give the IUPAC name of each molecule.*

Follow the general IUPAC rules for naming compounds.

a. $CH_3CH(CH_3)CH_2CH_2CH_3$
       2-methylpentane

b. $CH_3CH_2CH(CH_3)CH_2CH_3$
       3-methylpentane

76

c. $CH_3-CH_2-CH_2-CH_2-CH_2$
     $|$ (with CH_3 above the last carbon)

hexane

d. $CH_3-CH-CH_2-CH_3$
     $|$
     CH_2CH_3

3-methylpentane

4.71   *Draw and name six of the constitutional isomers with the formula* C_7H_{16}.

Start with the normal straight chain molecule and begin making isomers by
shortening the parent chain by one methyl group and placing that methyl group in
as many different places as possible. Next, take off two methyl groups and place
them on the remaining parent chain to additional structures. Continue the process
until you have created six structures. Remember that simply bending or turning
the molecule does not make it a different structure. For the molecule C_7H_{16}, there
are a total of nine constitutional isomers.

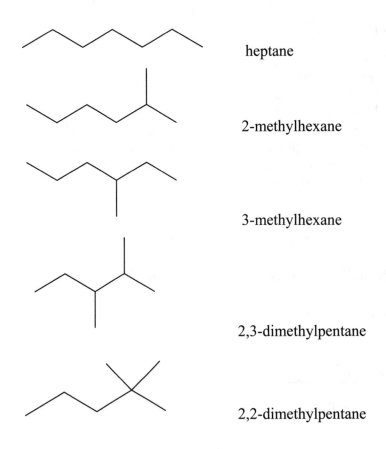

heptane

2-methylhexane

3-methylhexane

2,3-dimethylpentane

2,2-dimethylpentane

77

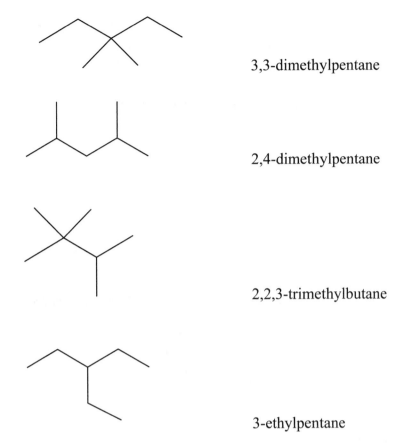

3,3-dimethylpentane

2,4-dimethylpentane

2,2,3-trimethylbutane

3-ethylpentane

4.73   *Which of the following pairs of molecules are constitutional isomers?*

Constitutional isomers are molecules that have the same molecular formula but whose atoms are connected differently. If the two molecules have the same molecular formula and the atoms are connected the same way, then they are identical molecules. If they have the same molecular formula but different atomic connections, then they are constitutional isomers. The skeletal structures are given to show the atomic connections.

a. CH₃CH₂CH₂CH₃ and CH₃CH(CH₃)CH₃

constitutional isomers

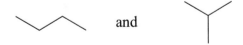   and

b. CH₃CH(CH₃)CH₂CH₃ and CH₃CH₂CH(CH₃)CH₃

identical molecules

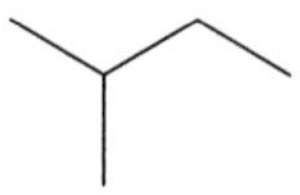

and

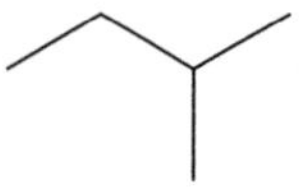

c. CH₃CH(CH₂CH₃)CH₂CH₃ and CH₃CH₂CH(CH₃)CH₂CH₃

c. $CH_3CH(CH_2CH_3)CH_2CH_3$ and $CH_3CH_2CH(CH_3)CH_2CH_3$

identical molecules

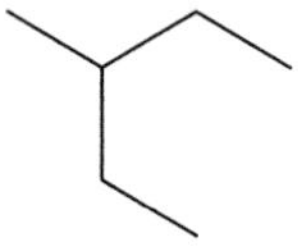

and

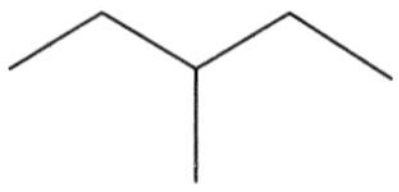

4.75   *Which pairs of molecules are constitutional isomers? Which are identical?*

a. $CH_3CH_2CH_2CH_2CH_2CH_3$ and CH_3CHCH_3 over $CH_3CH_2CH_3$

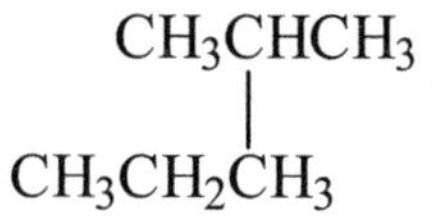

constitutional isomers

b. 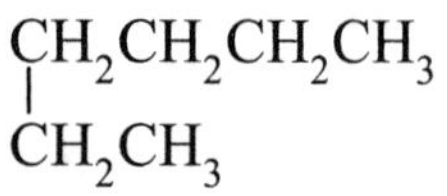 $CH_2CH_2CH_2CH_3$ / CH_2CH_3 and $CH_2CH_2CH_3$ / $CH_3CH_2CH_2$

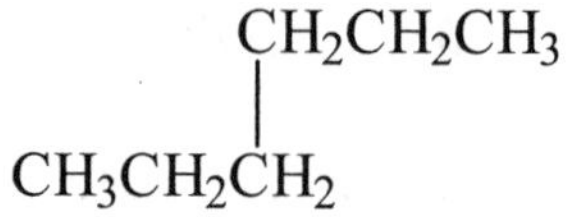

identical molecules

c. 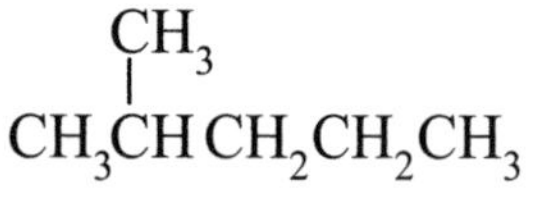 $CH_3CHCH_2CH_2CH_3$ with CH_3 and $CH_3CHCHCH_3$ with CH_3 and CH_3

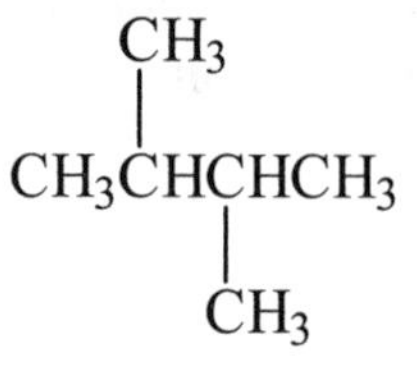

constitutional isomers

4.77   *Which are constitutional isomers?*

a. pentane and 2-methylpentane

No. They have different molecular formulas. Pentane is C_5H_{12} and 2-methylpentane is C_6H_{14}.

b. 2-methylpentane and 3-methylpentane

Yes. Both molecules have the same formula, C_6H_{14}. Different names indicate different atomic connections.

c. 2,2-dimethylpropane and pentane

Yes. Both molecules have the same formula, C_5H_{12} and different names (different atomic connections).

d. 2,2-dimethylpropane and cyclopentane

No. The molecules do not have the same molecular formula.

4.79 *Which pairs of molecules are constitutional isomers? Which are different conformations of the same molecule?*

To be constitutional isomers the molecules must have the same molecular formula (which all of the molecules in both a and b do); and they must have different atomic connections. The molecules in part a fail on the second criterion since both molecules are butane and have the same atomic connections. In part b, however, the first molecule is butane and the second is 2-methylpropane. This makes the molecules constitutional isomers. Since the molecules in part a are the same molecule with the hydrogens rotated into different special orientations, they are different conformations of the same molecule.

a.

Different conformations. The molecules have the same molecular formula and the same atomic connections. They differ only by the confirmation around the second carbon atom from the left.

b.

80

Constitutional isomers. The molecules have the same molecular formula but different atomic connections.

4.81   *a. How are constitutional isomers and conformations similar?*

In order to be constitutional isomers or conformations, molecules must have the same molecular formula.

b. How are constitutional isomers and conformations different?

Constitutional isomers have different atomic connections. Conformations have the same atomic connections, but different three-dimensional shapes that are interchanged by bond rotation.

4.83   *Give the IUPAC name for each molecule.*

a.

  methylcyclobutane
The parent ring is cyclobutane. The methyl group is named as a substituent. No numbering required because there is only one substituent.

b.

1,3-dimethylcylohexane
The parent ring is cyclohexane. There are two methyl substituent groups at carbon 1 and carbon 3.

c.

1-ethyl-2-methylcyclopentane
The parent ring is cyclopentane. The substituent groups are named and numbered in alphabetical order: ethyl at carbon 1 and methyl at carbon 2.

d.

1,1-dimethyl-2-propylcyclopropane
The parent ring is cyclopropane. There are 3 substituent groups: 2 methyl and 1 propyl. Name and number in alphabetical order.

4.85 *Which molecule(s) in Problem 4.83 can exist as* cis *and* trans *isomers?*

For a cycloalkane to have *cis* and *trans* isomers, it must have at least two substituents attached to different ring atoms and the substituents must have different spatial orientations (on the same face of the ring – *cis*; on opposite sides – *trans*).

a. No. There is only one substituent.
b. Yes. There are two methyl substituent groups on different carbon atoms of the ring.
c. Yes. There are two substituent groups, ethyl and methyl, on different carbon atoms of the ring.
d. No. The same molecule results regardless of whether the propyl group points up or down relative to the two identical methyl substituent groups on the neighboring ring carbon atom.

4.87 *Draw and name the three ethylmethylcyclobutane constitutional isomers.*

To be constitutional isomers, the molecules must have different atomic connections. They also have different IUPAC names.

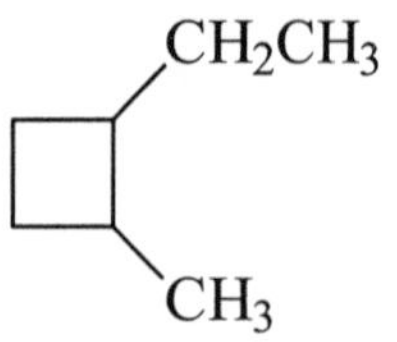

1-ethyl-2-methylcyclobutane

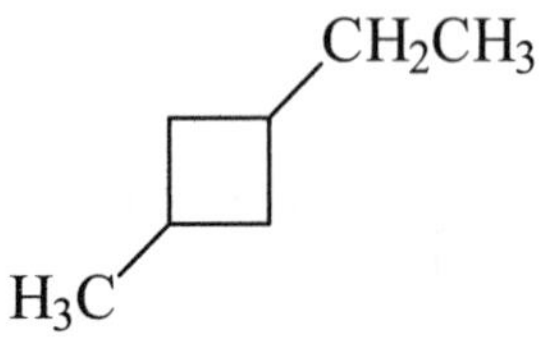

1-ethyl-3-methylcyclobutane

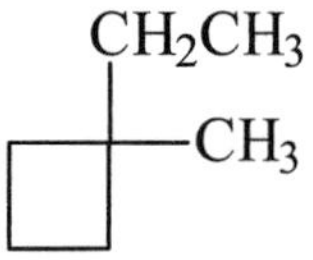

1-ethyl-1-methylcyclobutane

4.89 *a. What is the molecular formula of cyclopropane?*

C_3H_6. The expanded structure of cyclopropane is given below.

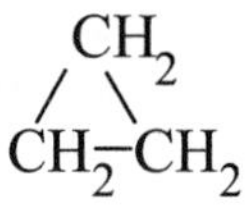

b. Draw a constitutional isomer of cyclopropane.

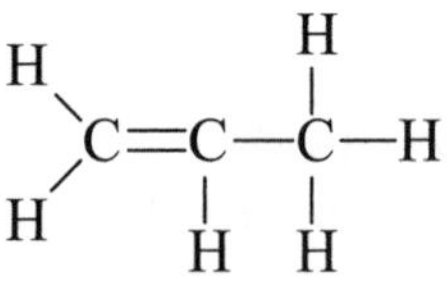

4.91 *Draw a side view of each cycloalkane.*

Trans means that substituents are to be drawn on opposite faces of the ring and *cis* means they are drawn on the same face.

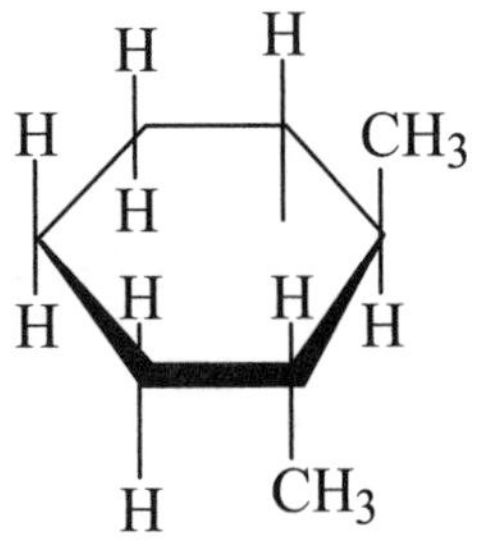

a. *trans-1,2-dimethylcyclohexane*

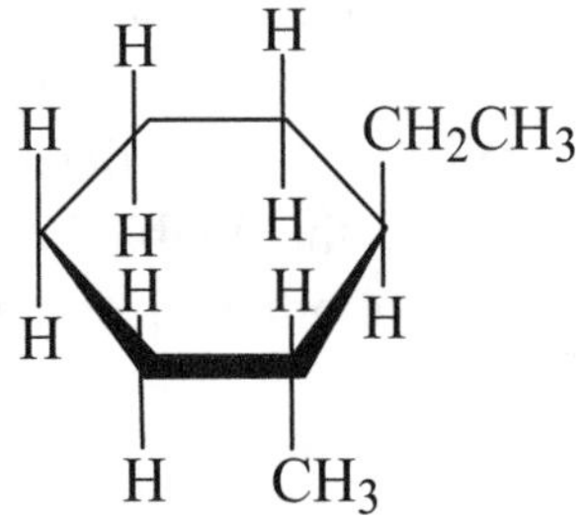

b. *trans-1-ethyl-2-methylcyclohexane*

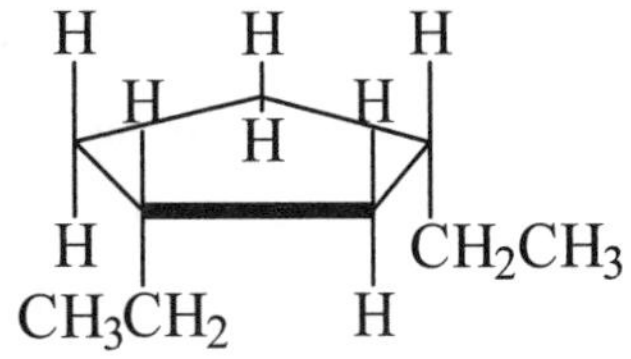

c. *cis-1,3-diethylcyclopentane*

4.93 *Give the complete IUPAC name (including the use of the term cis or trans) for each molecule.*

First identify the parent by counting the number of carbon atoms in the ring. Next, identify the substituent groups and assign numbering positions. Finally, use the prefix *cis* if the substituent groups are on the same face of the ring or *trans* if they are on opposite faces.

a.

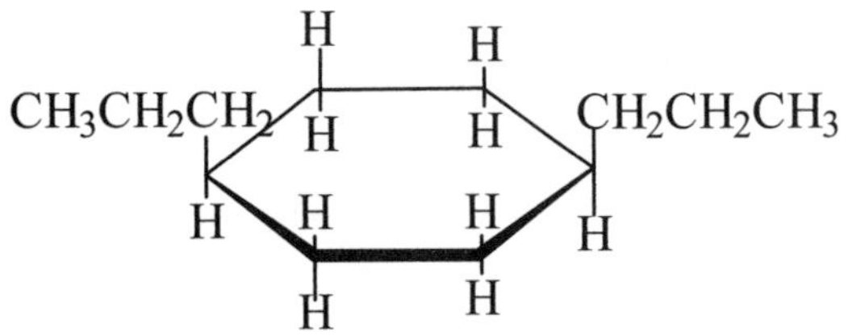

cis-1,4-dipropylcyclohexane

84

b.

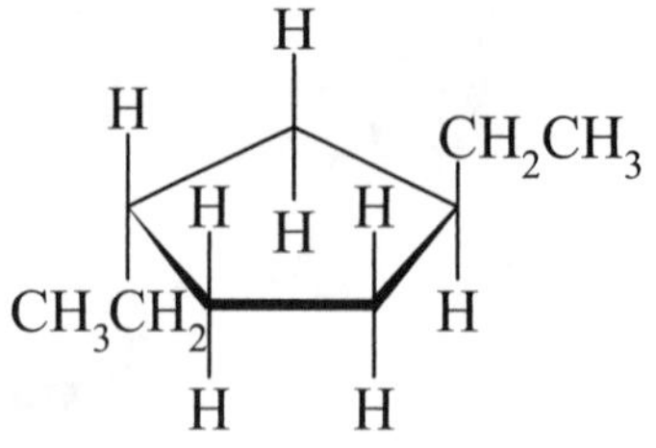
H
H
CH2CH3
H H H
CH3CH2
H
H H

trans-1,3-diethylcyclopentane

4.95 *Draw propene, showing the proper three-dimensional shape about each atom.*

The "ene" ending indicates a hydrocarbon with a double bond. The carbon atoms involved in the double bond are surrounded by a trigonal planar arrangement of atoms. The other carbon atom is surrounded by a tetrahedral arrangement.

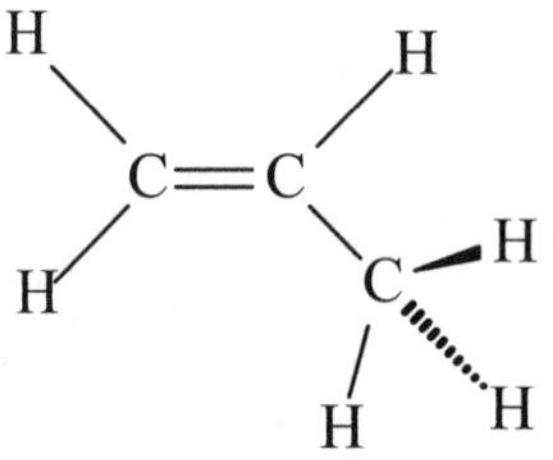
H
H
C=C
H
H
C
H
H

4.97 *Name each molecule.*

When naming hydrocarbons with double (alkene) or triple (alkyne) bonds, follow the same basic rules as for the alkanes, except for alkenes use an "ene" ending on the parent chain and for alkynes use an "yne" ending. The longest chain should include the double or triple bond. Number the longest chain from the end closest to the double or triple bond. Indicate the position of the double or triple bond using the lower number.

a. $CH_2{=}CHCH_2CH_3$

1-butene

b.

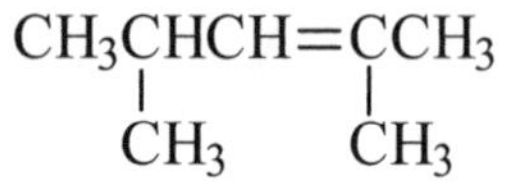
CH3CHCH=CCH3
CH3 CH3

2,4-dimethyl-2-pentene

CH₃C≡CCHCH₃
        |
        CH₃

c.

4-methyl-2-pentyne

4.99 *Draw each molecule*

Draw the parent chain first, placing the multiple bond in the position indicated in front of the parent chain name (remember "ene" means double bond and "yne" means triple bond). Next, starting from the same end used to count off the multiple bond position, count to the position number indicated and draw the substituents in their indicated positions.

a. *3-isopropyl-1-heptene*

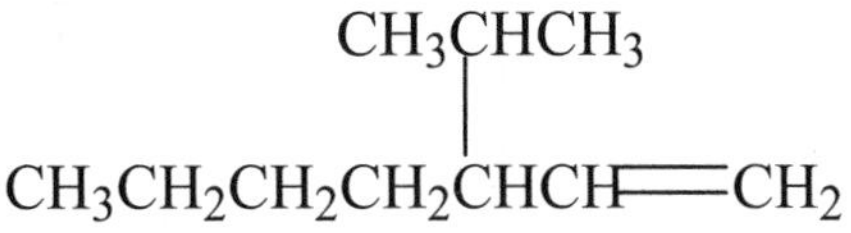

CH₃CHCH₃
      |
CH₃CH₂CH₂CH₂CHCH=CH₂

b. *2,3-dimethyl-2-butene*

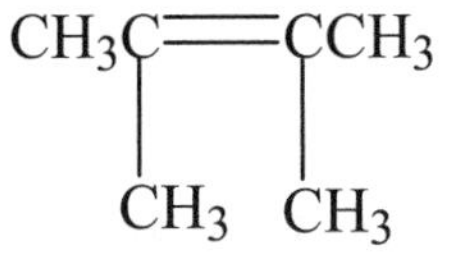

CH₃C=CCH₃
    | |
    CH₃ CH₃

c. *5-sec-butyl-3-nonyne*

CH₃CHCH₂CH₃
      |
CH₃CH₂CH₂CH₂CHC≡CCH₂CH₃

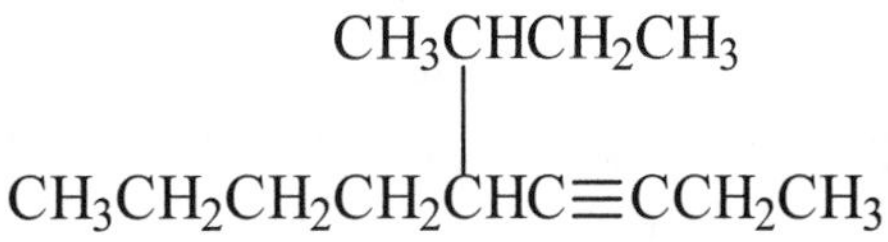

4.101 *The molecule shown is a termite trail marking pheromone. Which double bonds are cis and which are trans?*

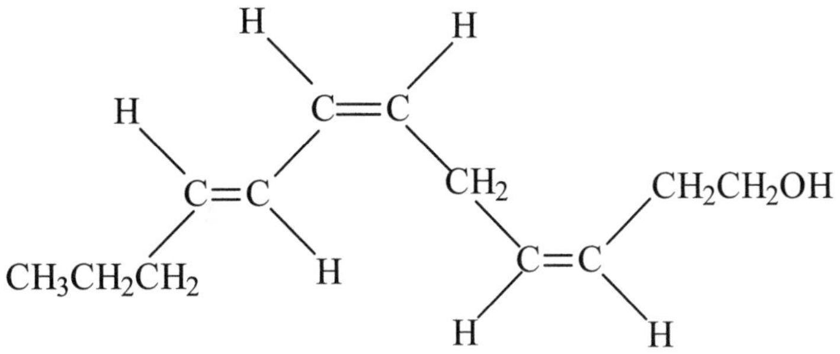

Recall that *cis* means that the atoms or groups of atoms being compared (in this case the attached carbon atoms) are on the same side of a line connecting the two double bonded carbon atoms, and *trans* means they are on opposite sides of that line.

Moving left to right across the molecule; the first double bond is *trans*; the second double bond is *cis*; and the third double bond is *cis*.

4.103   *Is the double bond in octyl methoxycinnamate (Figure 4.21) cis, trans, or neither?*

Trans, because the H atoms are on opposite sides of the C=C double bond.

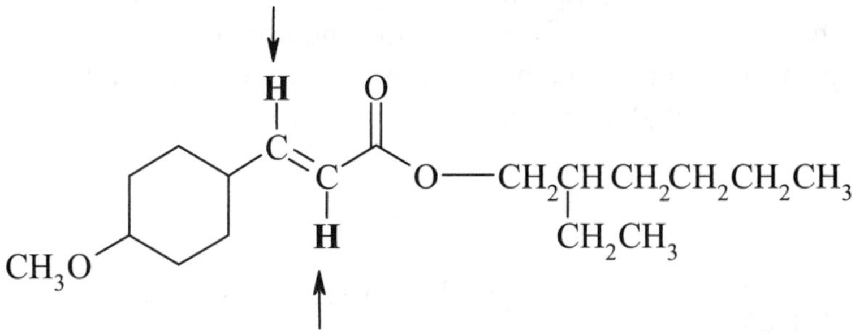

4.105   *Name each molecule.*

Aromatic rings follow the same naming rules as used earlier in the chapter for cycloalkanes. The substituent groups are numbered using the lowest set of digits. Substituent groups are named and numbered in alphabetical order. In the case of only two substituents, the *ortho* (side by side), *meta* (one carbon   between), *para* (on opposite side of the ring) designations may be used.

a.

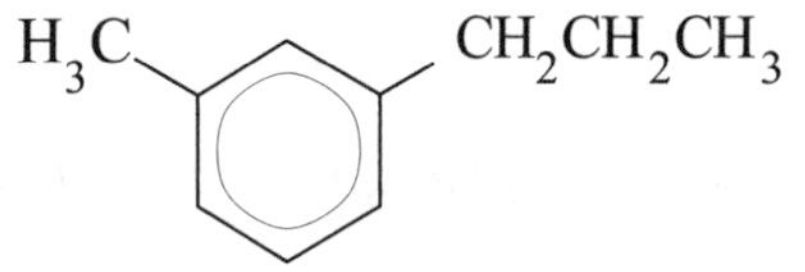

1-methyl-3-propylbenzene

b.

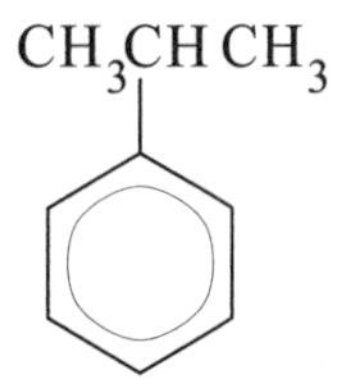

isopropylbenzene

c.

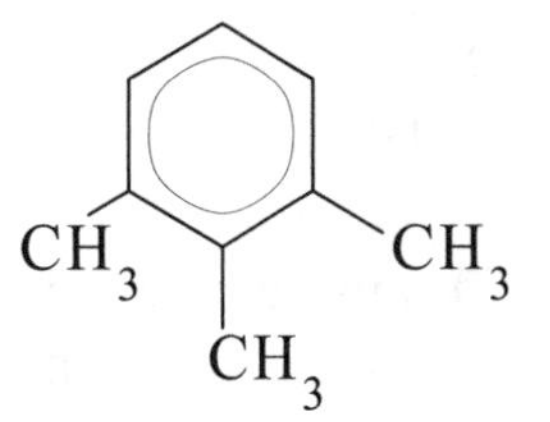

1,2,3-trimethylbenzene

4.107 *Draw each molecule.*

Draw the benzene ring first. Next, draw the substituents at the carbon number location designated in the name.

a. 1,3-dipropylbenzene

a.

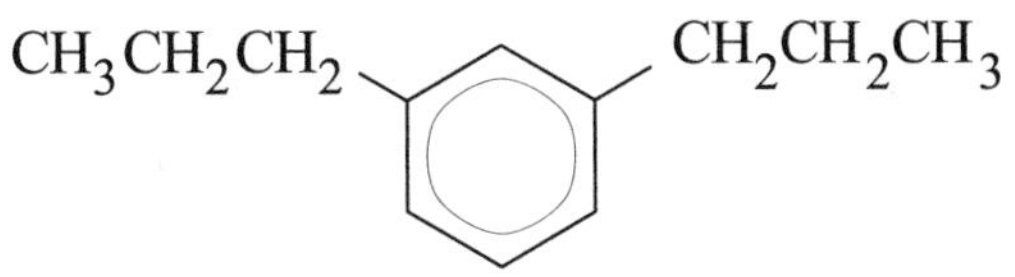

$CH_3CH_2CH_2$ —⬡— $CH_2CH_2CH_3$

b. p-diethylbenzene

b.

CH_2CH_3
⬡
CH_2CH_3

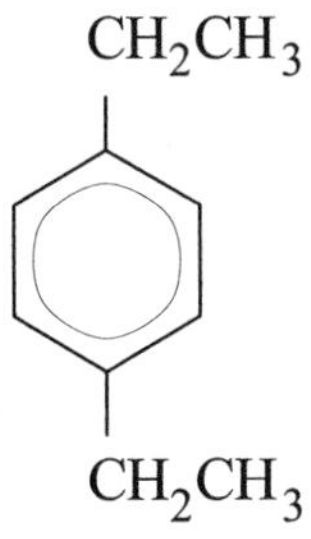

c. 4-isobutyl-1,2-dimethylbenzene

c.

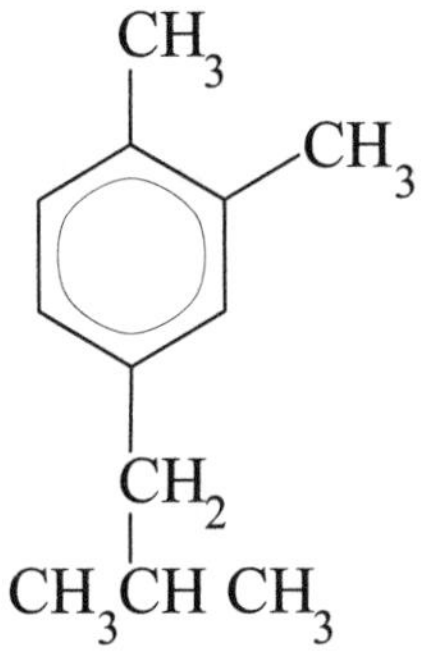

CH_3
CH_3
⬡
CH_2
$CH_3CH\ CH_3$

4.109 *Are covalent bonds broken when PrP^C is converted into PrP^{SC}? Explain.*

No, covalent bonds are not broken when PrP is converted into PrP^{SC}. The conversion occurs through free rotations around single bonds which change the overall shape of the PrP molecule.

4.111   *What properties are important for molecules used as sunscreens?*

The molecules should absorb UV-A and UV-B radiation and should not be toxic when applied to the skin.

4.115   *During ripening, bananas produce small amounts of ethylene. When bananas are shipped, why should they not be shipped in containers?*

The ethylene produced by bananas will build up inside the closed containers causing high levels of ethylene. This may cause the bananas to ripen rapidly, and possibly spoil.

4.117   *a. Draw methylcyclobutane.*

b. Draw a constitutional isomer of methylcyclobutane that has a ring and is a cis *geometric isomer.*

c. Draw a constitutional isomer of methylcyclobutane that has a ring, but has no geometric isomers.

d. Draw a constitutional isomer of methylcyclobutane that has no ring and is a cis *geometric isomer.*

e. Draw a constitutional isomer of methylcyclobutane that has no ring and has no geometric isomers.

$$CH_2=CH-CH_2-CH_2-CH_3$$

f. Name the major noncovalent force that attracts any one of the molecules in this problem to a similar molecule.

London force

4.119 *The term "organic" can have different meanings. What are two different ways to interpret a sign at the grocery store that reads "organic foods"?*

In one interpretation of "organic", the sign would be telling you that carbon-containing compounds are present. A second meaning is that the foods were grown utilizing nutrients that came from decaying organic matter.

Chapter 5
Reactions

Solutions to Problems

5.1   *The drawing below represents a chemical reaction.*

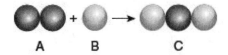

a. Write a balanced chemical equation for this reaction.

$A_2 + 4B \rightarrow 2C$

b. If 12 As are reacted with 40 Bs, what is the limiting reactant and what is the theoretical yield of C?

B is the limiting reactant and the theoretical yield is 20 C

The limiting reactant is the reactant that would get used up first. To determine which of the two reactants the limiting reactant is, calculate the amount of A_2 required to react with all of the B present using molar ratios from the balanced equation. If there is more than enough A_2 to consume all of B, then B is the limiting reactant because not all of the A_2 will be used up. If there is not enough A_2 to react with all of the B present, then A_2 is the limiting reactant because it will be used up first before all of the B present is used up.

$$40 \text{ B} \times \frac{1 \text{ A}_2}{4 \text{ B}} = 10 \text{ A}_2$$

Because only 10 A_2 are required to react with ALL of the B present, B will get used up first and there will be an excess of A_2.

To calculate the theoretical yield, use the amount of limiting reactant and the appropriate molar ratio to determine amount of C:

$$40 \text{ B} \times \frac{2 \text{ C}}{4 \text{ B}} = 20 \text{ C}$$

91

5.3   *Write the following sentence as a balanced chemical equation. Phosphorus reacts with chlorine (Cl₂) to produce phosphorus trichloride.*

$$2P + 3Cl_2 \rightarrow 2PCl_3$$

Prefixes like "tri" tell you the subscript for the element in the compound's formula. First translate the names to symbols: $P + Cl_2 \rightarrow PCl_3$. Next, check to see if the equation is balanced. It is not because there are 2 Cl atoms on the left and 3 Cl atoms on the right. Place the coefficient 2 in front of the PCl_3 ($2PCl_3$) which makes 6 Cl atoms on the right that can be balanced by placing a 3 in front of the Cl_2 ($3Cl_2$) to make 6 Cl atoms on the left. The $2PCl_3$ also means you have 2 P atoms on the right, so you need to place a 2 in front of the P on the left (2P) in order to have 2 P atoms on that side. This makes the equation balanced.

5.5   *What would you observe if you carried out the following reactions?*

a. $NH_3(g) + HCl(g) \rightarrow NH_4Cl(s)$

Formation of a solid.

The small letters given after the formulas indicate the physical state of the reactants and products: *(s)* means solid, *(l)* means liquid, *(g)* means gas, and *(aq)* means aqueous (dissolved in water). The *(s)* on the NH_4Cl *(s)* indicates it is produced as a solid and, therefore, you would observe a solid being produced during the reaction of the two gases. The solid is in the form of microscopic particles suspended in space.

b. $HCO_3^-(aq) + H_3O^+(aq) \rightarrow CO_2(g) + 2H_2O(l)$

Bubbling, indicating the formation of a gas.

The $CO_2(g)$ would be observed as a bubbling gas formed by mixing two liquid solutions.

5.7   *Balance the reaction equations.*

a. $SO_2 + O_2 \rightarrow SO_3$

$2SO_2 + O_2 \rightarrow 2SO_3$
You have 3 O atoms on the right, and 4 on the left. Place a 2 in front of SO_3, ($2SO_3$). This gives 6 O atoms and 2 S atoms on the right, so place a 2 in front of the SO_2 on the left, ($2SO_2$). This balances the S atoms and gives 4 O atoms to add to the O_2, for a total of 6 O atoms on the left.

b. $NO + O_2 \rightarrow NO_2$

$2NO + O_2 \rightarrow 2NO_2$
There are 3 O atoms on the left and 2 on the right. Place a 2 in front of the NO_2 ($2NO_2$). This means you need 2 in front of the NO, (2NO), so that the N atoms are balanced. This also gives 4 O atoms on both sides and makes the equation balanced.

5.9   *Balance the reaction equations.*

a. $K + Cl_2 \rightarrow KCl$

$2K + Cl_2 \rightarrow 2KCl$
There is 1 Cl atom on the right and 2 on the left. Place a 2 in front of the KCl (2KCl). This means that you need 2 in front of the K (2K) so that the K atoms are balanced. The equation is balanced.

b. $CH_4 + Cl_2 \rightarrow CH_2Cl_2 + HCl$

$CH_4 + 2Cl_2 \rightarrow CH_2Cl_2 + 2HCl$
There are 3 H atoms on the right and 4 on the left. Place a 2 in front of the HCl (2HCl). This means you need 2 in front of the Cl_2 ($2Cl_2$) so that the Cl atoms are balanced. The equation is balanced.

5.11   *Balance the reaction equations.*

Follow a similar process as in previous problems to balance these equations. In many cases, finding the balancing coefficients requires a trial-and-error method.

a. $Al + CuO \rightarrow Al_2O_3 + Cu$

$2Al + 3CuO \rightarrow Al_2O_3 + 3Cu$

b. $Mg + P_4 \rightarrow Mg_3P_2$

$6Mg + P_4 \rightarrow 2Mg_3P_2$

5.13   *Balance the reaction equations. It may help to think in terms of nitrate ion, ammonium ion, and sulfate ion, rather than individual N, O, H, and S atoms.*

a. $CaCl_2 + AgNO_3 \rightarrow AgCl + Ca(NO_3)_2$

$CaCl_2 + 2AgNO_3 \rightarrow 2AgCl + Ca(NO_3)_2$
As stated in the problem, treating the polyatomic ion NO_3^- as an "individual particle" helps to simplify the balancing process:

$CaCl_2 + AgNO_3$		$\rightarrow$	$AgCl + Ca(NO_3)_2$	
Ca	1		Ca	1
Cl	2		Cl	1
Ag	1		Ag	1
NO_3^-	1		NO_3^-	2

As is apparent from the table above, placing a 2 next to AgCl doubles the Cl on the product side to equal the number of Cl on the reactant side. This also doubles the Ag so a 2 needs to be placed next to $AgNO_3$ on the reactant side. This makes the number of Ag atoms equal on both sides (2) and also makes the number of NO_3^- ions equal on both sides (2). Using these two balancing coefficients results in the balanced equation given above.

b. $(NH_4)_2SO_4 + CaBr_2 \rightarrow NH_4Br + CaSO_4$

$(NH_4)_2SO_4 + CaBr_2 \rightarrow 2NH_4Br + CaSO_4$
Count each of the polyatomic ions as an individual particle (NH_4^+ and SO_4^{2-}). Proceed with a trial and error method to determine the balancing coefficients as was done in the previous problems.

5.15   *Balance the reaction equations.*

   a. Cl₂ + NaBr → NaCl + Br₂

   $Cl_2 + 2NaBr \rightarrow 2NaCl + Br_2$

   b. BiCl₃ + H₂O → Bi₂O₃ + HCl

   $2BiCl_3 + 3H_2O \rightarrow Bi_2O_3 + 6HCl$

5.17   *Classify each reaction as involving synthesis, decomposition, single replacement, or double replacement.*

   a. 2Na + Cl₂ → 2NaCl

   Synthesis. Two elements combine to form a compound.

   b. 2K + 2H₂O → H₂ + 2KOH

   Single replacement. K replaces one hydrogen atom in H_2O.

5.19   *Classify each reaction as involving synthesis, decomposition, single replacement, or double replacement.*

   a. HBr + NaOH → NaBr + H₂O

   Double replacement. Na and H exchange places to form two different compounds.

   b. Cu + 2AgNO3 → Cu(NO₃)₂ + 2Ag

   Single replacement. Cu replaces Ag in the compound.

   c. 2S + 3 O₂ → 2SO₃

   Synthesis. S and O_2 combine to form a compound.

d. $H_2SO_4 \rightarrow H_2O + SO_3$

Decomposition. A compound breaks down to form two simpler compounds.

5.21 *Classify the reactions in Problem 5.6 as involving synthesis, decomposition, single replacement or double displacement.*

a. $AgNO_3(aq) + KCl(aq) \rightarrow AgCl(s) + KNO_3(aq)$

Double replacement. Ag is replaced by K to form KNO_3 and K is replaced by Ag to form AgCl.

b. $Al(s) + Fe_2O_3(s) \rightarrow Fe(l) + Al_2O_3(s)$

Single replacement. Al replaces Fe in Fe_2O_3 to form Al_2O_3.

5.23 *Classify the reactions in Problem 5.11 as involving synthesis, decomposition, single replacement or double displacement.*

a. $Al + CuO \rightarrow Al_2O_3 + Cu$

Single replacement. Al replaces Cu in Cuo to form Al_2O_3.

b. $Mg + P_4 \rightarrow Mg_3P_2$

Synthesis. Mg and P_4 combine to form a compound, Mg_3P_2.

5.25 *The reaction shown here is one step in the fermentation process. To which reaction type does the reaction belong?*

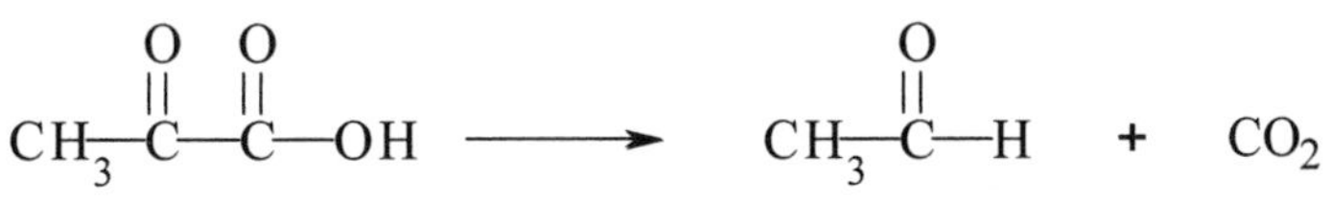

$$\underset{\text{Pyruvic acid}}{CH_3-\overset{\overset{O}{\|}}{C}-\overset{\overset{O}{\|}}{C}-OH} \longrightarrow \underset{\text{Acetaldehyde}}{CH_3-\overset{\overset{O}{\|}}{C}-H} + CO_2$$

Decomposition. Pyruvic acid decomposes into two simpler compounds, acetaldehyde and carbon dioxide.

5.27 *As we will see in Section 8.4, an ester can be prepared in the following way. Does the reaction involve synthesis, decomposition, single replacement, or double displacement?*

$$CH_3-\overset{\overset{O}{\|}}{C}-\boxed{OH} + H-\boxed{OCH_3} \xrightarrow{H^+} CH_3-\overset{\overset{O}{\|}}{C}-O-CH_3 + HOH$$

Double replacement. $-OCH_3$ in $HOCH_3$ replaces the $-OH$ in the first reactant molecule to form the first product molecule. The $-OH$ replaces $-OCH_3$ in $HOCH_3$ to form HOH.

5.29 *Draw the products of each hydrolysis reaction.*

When an ester is hydrolyzed in the presence of H^+, the two products are a carboxylic acid and an alcohol.

a.
$$H-\overset{\overset{O}{\|}}{C}-O-CH_2-CH_3 + H_2O \xrightarrow{H^+}$$

$$\boxed{H-\overset{\overset{O}{\|}}{C}-OH + HOCH_2CH_3}$$

b.
$$CH_3-\overset{\overset{O}{\|}}{C}-O-CH_2-\bigcirc + H_2O \xrightarrow{H^+}$$

$$\boxed{CH_3-\overset{\overset{O}{\|}}{C}-OH + HOCH_2-\bigcirc}$$

c.
$$CH_3-CH_2-\overset{\overset{O}{\|}}{C}-O-CH_2-CH_2-CH_3 + H_2O \xrightarrow{H^+}$$

97

$$CH_3CH_2-\overset{\displaystyle O}{\overset{\|}{C}}-OH \quad + \quad HOCH_2CH_2CH_3$$

5.31 *Draw the missing reactant for each hydrolysis reaction.*

In each case, the missing reactant is an ester hydrolyzed into a carboxylic acid and an alcohol. The components of the ester molecule can be derived from the carboxylic acid molecule and the alcohol molecule as shown in part a.

a.
$$? \quad + \quad H_2O \quad \xrightarrow{H^+} \quad CH_3CH_2-\overset{\displaystyle O}{\overset{\|}{C}}-OH \quad + \quad HOCH_3$$

The missing reactant is
$$CH_3CH_2-\overset{\displaystyle O}{\overset{\|}{C}}-O-CH_3 \quad .$$

b.
$$? \quad + \quad H_2O \quad \xrightarrow{H^+} \quad H-\overset{\displaystyle O}{\overset{\|}{C}}-OH \quad + \quad HO-CH_2-\underset{\underset{\displaystyle CH_3}{|}}{CH}-CH_3$$

The missing reactant is
$$H-\overset{\displaystyle O}{\overset{\|}{C}}-O-CH_2-\underset{\underset{\displaystyle CH_3}{|}}{CH}-CH_3$$

c.
$$? \quad + \quad H_2O \quad \xrightarrow{H^+} \quad C_6H_5-\overset{\displaystyle O}{\overset{\|}{C}}-OH \quad + \quad HOCH_3$$

The missing reactant is
$$C_6H_5-\overset{\displaystyle O}{\overset{\|}{C}}-O-CH_3$$

5.33 *Draw the hydration product formed when each alkene is reacted with H_2O in the presence of H^+.*

a. trans-3-hexene

Hydration of an alkene is the process of adding a water molecule across a double bond. The end result is that the double bond is replaced by a H atom on one of the originally double bonded carbon atoms and an OH group is placed on the other. To see the product that will form, draw the structure of the reactant molecule, remove one of the bonds in the double bond and place hydrogen on one side and an OH on the other.

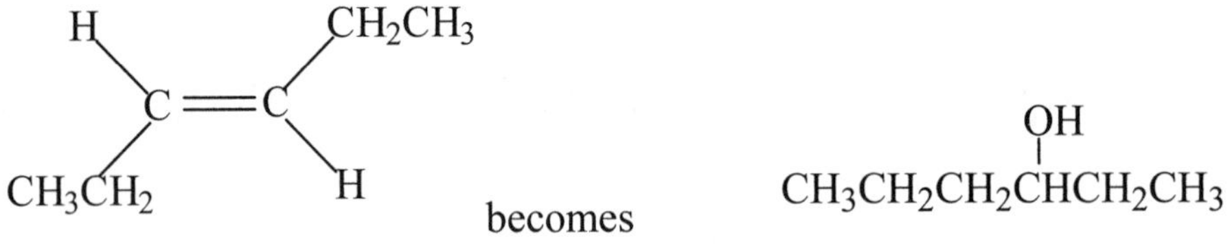

becomes $CH_3CH_2CH_2\overset{\overset{\displaystyle OH}{|}}{C}HCH_2CH_3$

b. cis-3-hexene

See directions in part a.

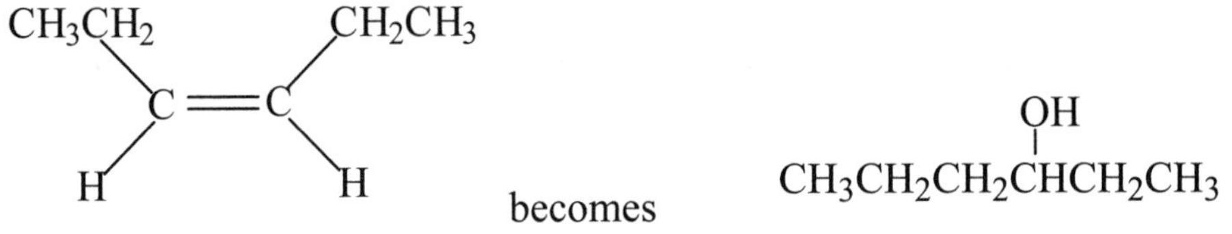

becomes $CH_3CH_2CH_2\overset{\overset{\displaystyle OH}{|}}{C}HCH_2CH_3$

c. 1,2-dimethylcyclopentene

See directions given in part a.

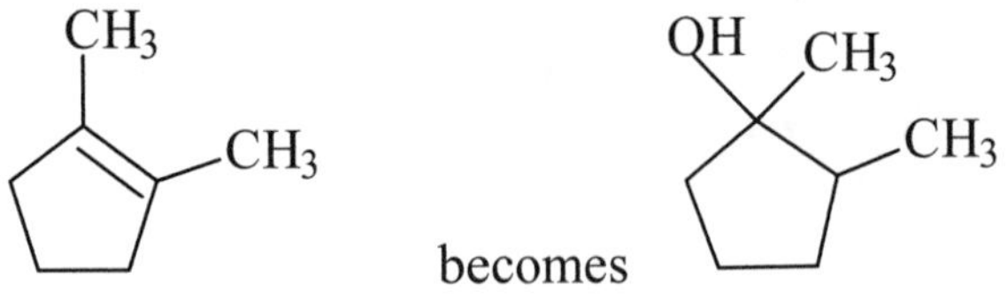

becomes

5.35 *Draw the missing reactant for each hydration reaction.*

99

In each case, the missing reactant is an alkene. The hydration of an alkene produces an alcohol. In a hydration, a water molecule is split apart with a H atom making a bond to one of the C atoms in a C=C and the –OH group forming a bond with the other C atom in the C=C group as shown for part a below.

a.
$$? \quad + \quad H_2O \quad \xrightarrow{H^+} \quad CH_3CH_2OH$$

The missing reactant is
$$\begin{array}{c} H-C=C-H \\ \quad | \quad | \\ \quad H \quad H \end{array}$$

$$\begin{array}{c} H-C=C-H \\ | \quad | \\ H \quad H \end{array} \quad + \quad \boxed{H\,OH} \quad \xrightarrow{H^+} \quad CH_3CH_2OH$$

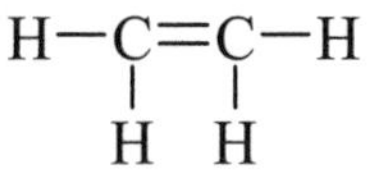

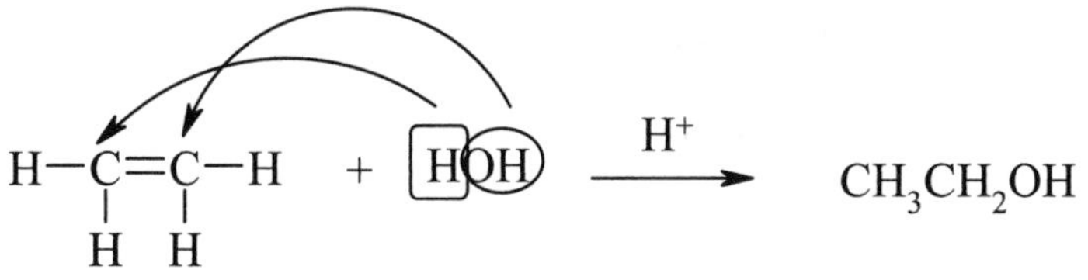

b.
$$? \quad + \quad H_2O \quad \xrightarrow{H^+} \quad \begin{array}{c} OH \\ | \\ CH_3-CH-CH_3 \end{array}$$

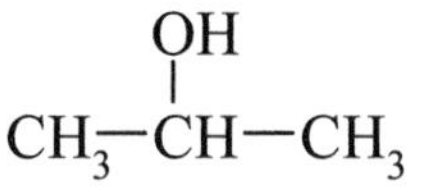

The missing reactant is
$$\begin{array}{c} H-C=C-CH_3 \\ \quad | \quad | \\ \quad H \quad H \end{array}$$

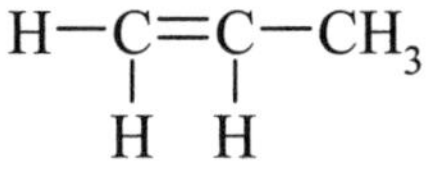

c.
$$? \quad + \quad H_2O \quad \xrightarrow{H^+} \quad$$

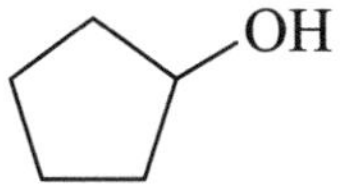

The missing reactant is

100

5.37   *Draw the organic dehydration product formed when each alcohol is heated in the presence of H^+.*

Dehydration is the reverse of hydration. First, identify the location of an H atom and OH group that are on side-by-side carbons. These are removed and replaced with another bond line between the two carbons creating a double bond. The H and OH combine to give water as a second product.

a.

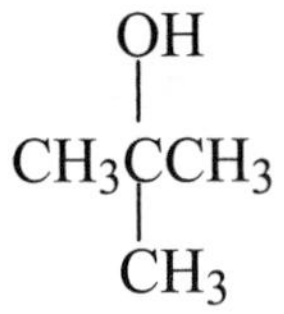

$$CH_3\underset{\underset{CH_3}{|}}{\overset{\overset{OH}{|}}{C}}CH_3 \qquad becomes \qquad CH_3\underset{\underset{CH_3}{|}}{C}{=}CH_2$$

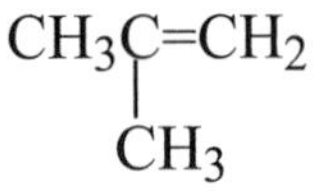

b.

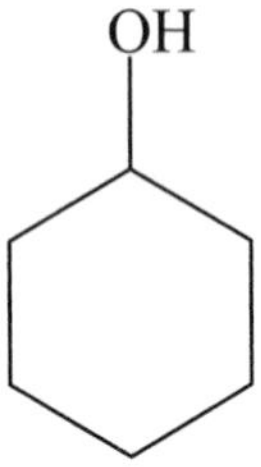

becomes

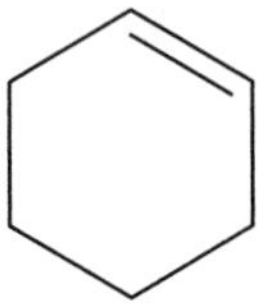

c.

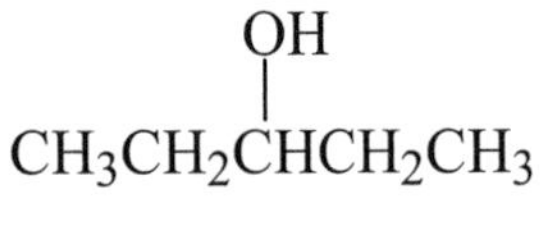

$$CH_3CH_2\underset{\overset{OH}{|}}{C}HCH_2CH_3 \quad becomes \quad CH_3CH_2CH{=}CHCH_3$$

5.39   *Draw each missing reactant.*

In each case, determine the missing reactant based on the products.

a. $? \; + \; H_2O \; \xrightarrow{H^+} \;$

101

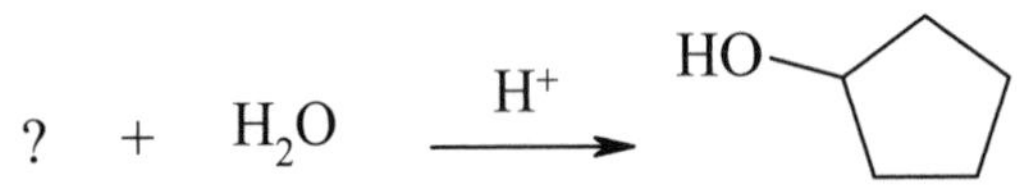

The missing reactant is

This is a hydration reaction of an alkene because the product is an alcohol.

b. 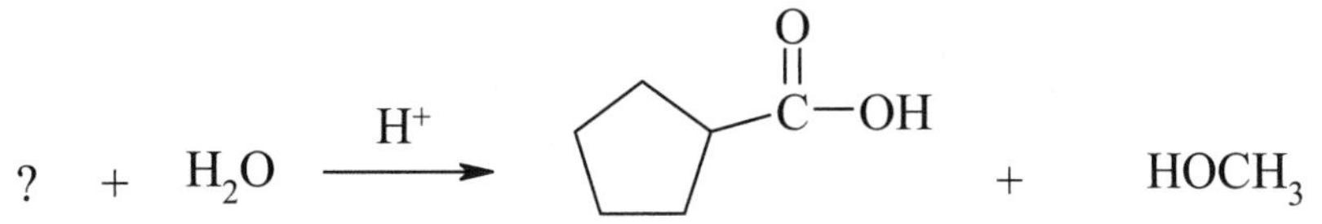 ? + H₂O →(H⁺)

+ HOCH₃

The missing reactant is

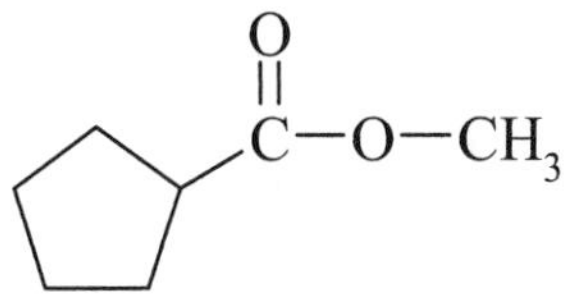

The reaction is the hydrolysis of an ester because the products are a carboxylic acid and an alcohol.

c. 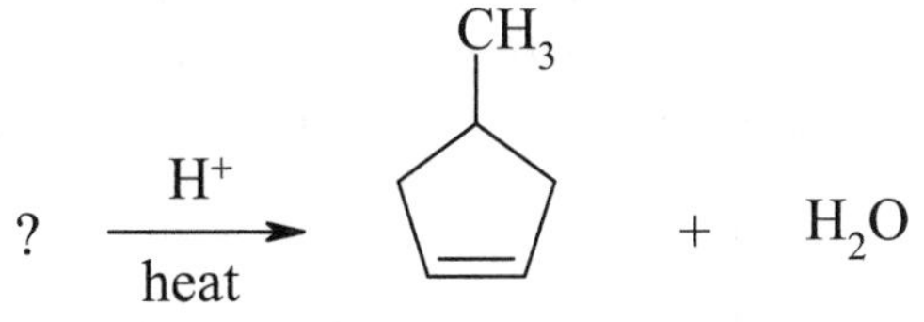 ? →(H⁺, heat)

+ H₂O

The missing reactant is

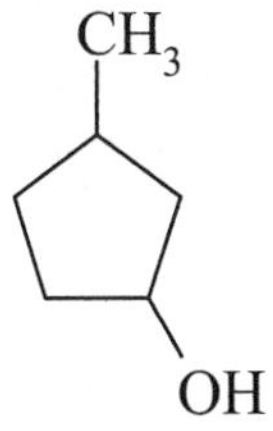

This is a dehydration reaction of an alcohol producing an alkene and water.

5.41 *Aspirin (acetylsalicylic acid) is an ester of acetic acid.*

$$
\begin{array}{c}
\text{O} \\
\parallel \\
\text{HO}-\text{C} \qquad \begin{array}{c}\text{O}\\ \parallel\end{array} \\
\text{(benzene ring)} \quad \text{O}-\text{C}-\text{CH}_3
\end{array}
$$

a. Circle the ester group in aspirin.

$$
\begin{array}{c}
\text{O} \\
\parallel \\
\text{HO}-\text{C} \\
\text{(benzene ring)} \quad \boxed{\text{O}-\text{C}-}\text{CH}_3
\end{array}
$$

b. Aspirin is sold with cotton placed in the neck of the bottle to help keep moisture out. When exposed to water, the ester group of aspirin slowly hydrolyzes. Draw the hydrolysis products obtained if aspirin is reacted with H_2O in the presence of H^+.

Hydrolysis of an ester in the presence of H^+ results in the formation of the carboxylic acid molecule and an alcohol molecule. In this case, the acetyl group becomes part of the carboxylic acid molecule and the salicylic acid group becomes the alcohol molecule.

$$
\begin{array}{c}
\text{O} \\
\parallel \\
\text{C}-\text{OH} \\
\text{(benzene ring)}\;\text{OH}
\end{array}
\quad + \quad
\begin{array}{c}
\text{O} \\
\parallel \\
\text{HO}-\text{C}-\text{CH}_3
\end{array}
$$

5.43   *The ester shown below is reacted with H_2O and H^+ to form carboxylic acid A and alcohol B. Alcohol B is heated in the presence of H^+ to produce molecule C. Draw A, B, and C.*

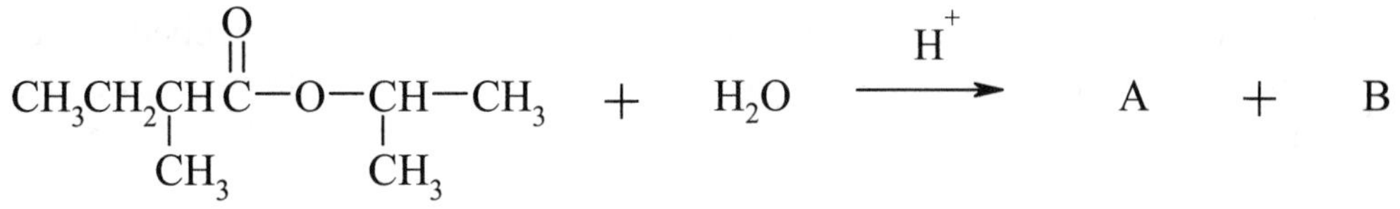

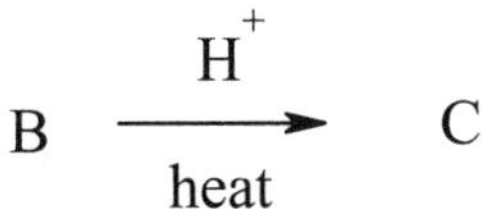

$$
\begin{array}{c}
\underset{\underset{CH_3}{|}}{CH_3CH_2CH}\overset{\overset{O}{\|}}{C}-O-\underset{\underset{CH_3}{|}}{CH}-CH_3 \quad + \quad H_2O \quad \xrightarrow{\;H^+\;} \quad A \quad + \quad B
\end{array}
$$

$$
B \xrightarrow[\text{heat}]{\;H^+\;} C
$$

When an ester is reacted with H_2O and H^+, it undergoes a hydrolysis reaction producing a carboxylic acid (molecule A) and an alcohol (molecule B):

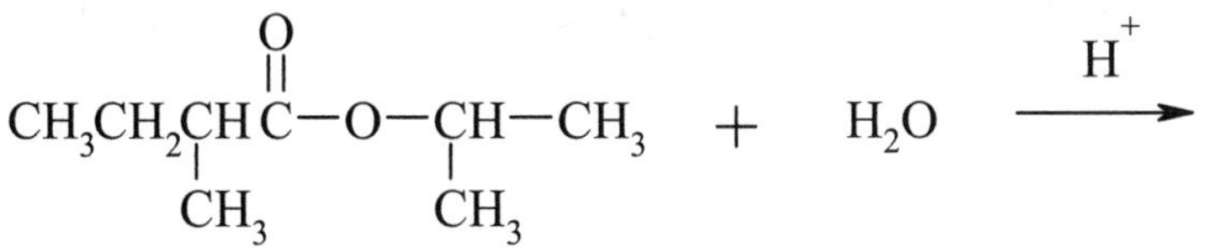

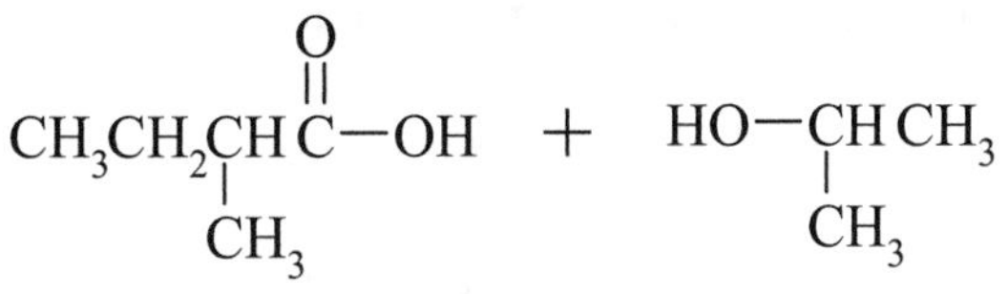

$$
\begin{array}{c}
\underset{\underset{CH_3}{|}}{CH_3CH_2CH}\overset{\overset{O}{\|}}{C}-O-\underset{\underset{CH_3}{|}}{CH}-CH_3 \quad + \quad H_2O \quad \xrightarrow{\;H^+\;}
\end{array}
$$

$$
\underset{\underset{CH_3}{|}}{CH_3CH_2CH}\overset{\overset{O}{\|}}{C}-OH \quad + \quad HO-\underset{\underset{CH_3}{|}}{CH}CH_3
$$

     molecule A   molecule B

When alcohol B is heated in the presence of H^+, it undergoes a dehydration reaction to form an alkene (molecule C).

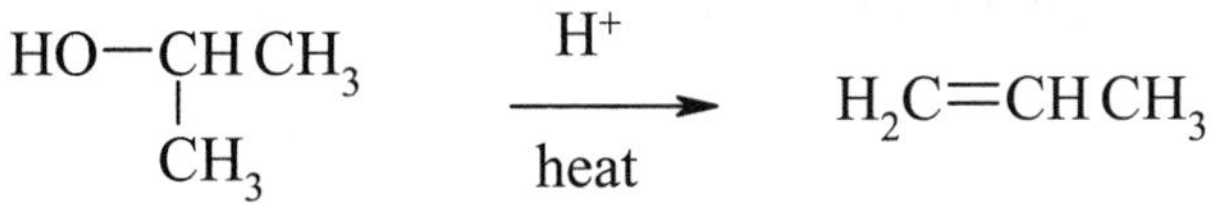

$$
HO-\underset{\underset{CH_3}{|}}{CH}CH_3 \quad \xrightarrow[\text{heat}]{\;H^+\;} \quad H_2C=CHCH_3
$$

molecule B   molecule C

104

5.45   *Zinc reacts with copper(II) sulfate according to the equation*
$$Zn(s) \ + \ CuSO_4(aq) \ \rightarrow \ ZnSO_4(aq) \ + \ Cu(s)$$

In an oxidation and reduction reaction, the atom or ion that loses electrons is oxidized and the atom or ion that gains electrons is reduced. The chemical that receives electrons is the oxidizing agent and the chemical that donates electrons is the reducing agent.

a.   Is zinc oxidized or is it reduced?

Oxidized. Zinc is oxidized because it lost electrons, going from neutral Zn(s) to Zn^{2+}.

b.   Is copper(II) ion oxidized or is it reduced?

Reduced. The copper(II) ion is reduced because it gained electrons, going from a positively charged species, Cu^{2+}, to neutral Cu.

c.   What is the oxidizing agent?

Cu(II) ion is the oxidizing agent because it received electrons.

d.   What is the reducing agent?

Zn is the reducing agent because it donated electrons.

5.47   *Sodium metal reacts with oxygen gas (O_2) to form sodium oxide.*

a. Write a balanced equation for this oxidation-reduction reaction. (In this problem you need not worry about the physical state of the reactants or product.)

$$4 \, Na \ + \ O_2 \ \rightarrow \ 2 \, Na_2O$$

b. Which reactant is oxidized?

Na is oxidized because it lost electrons, going from neutral Na to Na^+.

c. Which reactant is reduced?

O_2 is reduced because it gained electrons, going from neutral O to O^{2-}.

d. What is the oxidizing agent?

O_2 is the oxidizing agent because it removed electrons.

e. What is the reducing agent?

Na is the reducing agent because it gave electrons.

5.49 *Classify the reactions in Problems 5.45 and 5.47 as involving synthesis, decomposition, single replacement, or double replacement.*

Problem 5.45: $Zn(s) + CuSO_4(aq) \rightarrow ZnSO_4(aq) + Cu(s)$
Single replacement.

Problem 5.47: $4\,Na + O_2 \rightarrow 2\,Na_2O$
Synthesis.

5.51 *A reaction of iron metal with oxygen gas produces ferrous oxide (FeO).*

$$Fe + O_2 \rightarrow FeO$$

a. Balance the equation.

There is 1 O atom on the right and 2 on the left. Place a 2 in front of the FeO, (2FeO). This gives two O atoms on the right. Since this also makes two Fe atoms on the right, place a 2 in front of the Fe (2Fe) and the equation is balanced.

$$2Fe + O_2 \rightarrow 2FeO$$

b. When FeO forms, has Fe been oxidized or has it been reduced?

Since oxidized means loss of electrons and reduced means gain of electrons, to determine which atom is oxidized or reduced, the charge on each atom or ion must be determined. Some of these can be predicted from the periodic table. When that is not possible (transition metals, for example) the charge is calculated using the other ions of known charge.

charges: 0   0   2+ 2-
$$2Fe + O_2 \rightarrow 2FeO$$

Fe is changing from a 0 charge to a 2+ charge which means it has lost 2 electrons so it oxidized.

Oxidized

c. *When FeO forms, has O_2 been oxidized or has it been reduced?*

Using the charges calculated in part b, the O_2 with 0 charge on each atom is changed to O^{2-} which means it has gained 2 electrons. It is reduced.

Reduced

5.53   *Ethanol (CH_3CH_2OH) is mixed with gasoline to produce gasohol, a cleaner burning fuel than gasoline. Write the balanced chemical equation for the complete oxidation of ethanol by O_2 to produce CO_2 and H_2O.*

The balanced equation is: $CH_3CH_2OH + 3O_2 \rightarrow 2CO_2 + 3H_2O$

See the steps below for how to arrive at this answer.

Write the formulas in equation format by placing the reactants on the left side and the products on the right side of the arrow.

$CH_3CH_2OH + O_2 \rightarrow CO_2 + H_2O$

To begin the balancing process, examine the equation carefully to see if it may already be balanced. In this case it is not. Next, note that there are more hydrogen atoms than any other element so this is probably a good element to start with. Since there are 6 H atoms on the left, place a 3 in front of H_2O ($3H_2O$). This makes six hydrogen atoms on both sides. Then, since there are 2 C atoms on the left, place a 2 in front of the CO_2 ($2CO_2$). This leaves only the O atoms to check. On the right there are 4 O atoms from the $2CO_2$ and 3 from the $3H_2O$ for a total of 7 O atoms on the right. On the left there is 1 O atom from the CH_3CH_2OH and 2 O atoms from the O_2. Complete the balancing by placing a 3 in front of the O_2 ($3O_2$). This makes $1 + 6 = 7$ O atoms on the left which balances the equation.

As recommended by the text, this equation can also be balanced by balancing C first, then H, then O.

5.57 *The chemical equation on page 164 shows the conversion of phenylalanine into tyrosine. The balanced equation for this reaction includes O_2 and a compound called tetrahydrobiopterin, $[BH_4]$, which reacts to form 4-α-hydroxybiopterin, $[BH_3OH]$.*

$$\text{phenylalanine} + O_2 + BH_4 \xrightarrow{\text{phenylalanine hydroxylase}}$$

$$\text{tyrosine} + BH_3OH$$

a. Is phenylalanine oxidized or is it reduced?

Oxidized. A new bond to an oxygen atom is formed and a bond to a H atom is lost.

b. Is tetrahydrobiopterin oxidized or reduced?

Oxidized. A new bond to an oxygen atom is formed and a bond to a H atom is lost.

c. Is O_2 oxidized or reduced?

Reduced. For each O atom, two bonds to H atoms are formed and a bond to another oxygen atom is lost.

5.59 *Draw the missing product of each reaction.*

Each of these reactions is a hydrogenation reaction as indicated by the presence of the H_2 reactant and the Pt catalyst. Each of the 2 H atoms of H_2 adds across the double bond converting the alkene to an alkane.

$$a. \quad CH_2{=}CHCH_2CH_3 \ + \ H_2 \xrightarrow{\text{Pt}} \ ?$$

The missing product is $CH_3CH_2CH_2CH_3$

$$CH_3-CH=CHCH_2CH_3 \quad + \quad H_2 \quad \xrightarrow{Pt} \quad ?$$
$$\overset{|}{CH_3}$$
b.

$$CH_3-CH-CH_2-CH_2-CH_3$$
$$\overset{|}{CH_3}$$
The missing product is

 $-CH_3 \quad + \quad H_2 \quad \xrightarrow{Pt} \quad ?$

c.

The missing product is

5.61 *Draw the skeletal structure of the product formed when each alkene is reacted with H_2, in the presence of Pt.*

a. cyclopentene

First, draw the structure of cyclopentene.

Upon reaction with H_2, the double bond is converted to a single bond. Note that the hydrogen atoms added will not be shown since you are drawing a skeletal structure.

b. 2-ethyl-3-methyl-1-pentene

109

First, draw the structure of 2-ethyl-3-methyl-1-pentene.

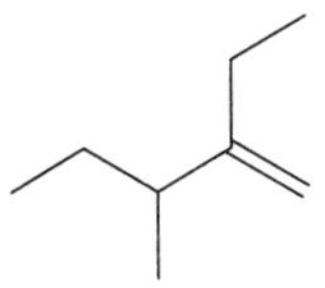

Upon reaction with H_2, the double bond is converted to a single bond. If you draw the answer as a skeletal structure, as is done here, the hydrogen atoms added will not be shown.

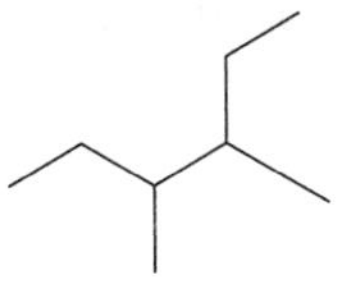

5.63 *Draw the saturated product expected when 3 mol of H_2, in the presence of Pt, are reacted with 1 mol of the termite trail marking pheromone shown below.*

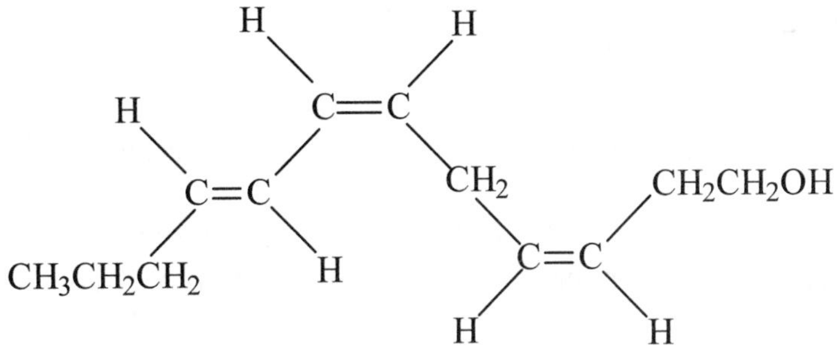

$CH_3(CH_2)_{10}CH_2OH$ is the saturated product expected.
3 mol of H_2 will replace all of the double bonds in 1 mol of the pheromone.

5.65 *For the reaction between carbon and oxygen to form carbon monoxide, beginning with 0.65 mol of C,*

$$2C(s) + O_2(g) \rightarrow 2CO(g)$$

a. how many moles of O_2 are required to completely consume the C?

Use the mole to mole ratio between C and O_2 given by the equation as a conversion factor to convert moles of C to moles of O_2.

$$0.65 \; \text{mol C} \; \times \; \frac{1 \; \text{mol } O_2}{2 \; \text{mol C}} = 0.33 \; \text{mol } O_2$$

b. how many moles of CO are obtained when the C is completely reacted?

Use the mole to mole ratio between C and CO given by the equation as a conversion factor to convert moles of C to moles of CO.

$$0.65 \; \cancel{mol \; C} \; \times \; \frac{2 \; mol \; CO}{2 \; \cancel{mol \; C}} \; = \; 0.65 \; mol \; CO$$

5.67   *For the combustion of methane, beginning with 3.15 mol of CH₄,*

$$CH_4(g) + 2O_2(g) \rightarrow CO_2(g) + 2H_2O(g)$$

a. How many moles of O₂ are required to totally consume the CH₄?

Use the mole to mole ratio for CH₄ to O₂ given by the equation as a conversion factor to convert moles of CH₄ to moles of O₂.

$$3.15 \; \cancel{mol \; CH_4} \; \times \; \frac{2 \; mol \; O_2}{1 \; \cancel{mol \; CH_4}} \; = \; 6.30 \; mol \; O_2$$

b. How many moles of CO₂ are obtained when the CH₄ is totally consumed?

Use the mole to mole ratio for CH₄ to CO₂ given by the equation as a conversion factor to convert from moles of CH₄ to moles of CO₂.

$$3.15 \; \cancel{mol \; CH_4} \; \times \; \frac{1 \; mol \; CO_2}{1 \; \cancel{mol \; CH_4}} \; = \; 3.15 \; mol \; CO_2$$

c. How many moles of H₂O are obtained when the CH₄ is totally consumed?

Use the mole to mole ratio for CH₄ to H₂O given by the equation as a conversion factor to convert from moles of CH₄ to moles of H₂O.

$$3.15 \; \cancel{mol \; CH_4} \; \times \; \frac{2 \; mol \; H_2O}{1 \; \cancel{mol \; CH_4}} \; = \; 6.30 \; mol \; H_2O$$

5.69 *For the reaction in Problem 5.51,*

a. how many grams of Fe are required to react completely with 4.22 x 10^{-3} mol of O_2?

The balanced equation for the reaction in Problem 5.51 is:

$$2Fe \ + \ O_2 \ \rightarrow \ 2FeO$$

First, convert the mol of O_2 to mol Fe using the molar ratio from the balanced equation. Then, use the molar mass of Fe to convert mol Fe to grams of Fe.

$$4.22 \ x \ 10^{-3} \ \text{mol } O_2 \ \text{ x } \ \frac{2 \ \text{mol Fe}}{1 \ \text{mol } O_2} \ \text{ x } \ \frac{55.85 \ \text{g Fe}}{1 \ \text{mol Fe}} \ = \ 0.471 \ \text{g Fe}$$

b. how many grams of FeO are produced from the complete reaction of 17.44g of Fe?

First, convert the grams of Fe to mol Fe and then, use the molar ratio between Fe and FeO to find mol FeO. Use the molar mass of FeO to convert to grams of FeO.

$$17.44 \ \text{g Fe} \ \text{ x } \ \frac{1 \ \text{mol Fe}}{55.85 \ \text{g Fe}} \ \text{ x } \ \frac{2 \ \text{mol FeO}}{2 \ \text{mol Fe}} \ \text{ x } \ \frac{71.85 \ \text{g FeO}}{1 \ \text{mol FeO}} \ = \ 22.44 \ \text{g FeO}$$

5.71 *Isopropyl alcohol (rubbing alcohol) can be produced by the reaction:*

$$CH_3CH=CH_2 \ + \ H_2O \ \xrightarrow{H^+} \ CH_3\overset{\overset{\displaystyle OH}{|}}{CH}CH_3$$

   propene   isopropyl alcohol

a. How many moles of H_2O, are required to completely react with 55.7 mol of propene?

Use the mole to mole ratio for C_3H_6 to H_2O given by the equation as a conversion factor to convert from moles of C_3H_6 to moles of H_2O.

$$55.7 \ \text{mol } C_3H_6 \ \text{ x } \ \frac{1 \ \text{mol } H_2O}{1 \ \text{mol } C_3H_6} \ = \ 55.7 \ \text{mol } H_2O$$

b. How many moles of H_2O are required to completely react with 1.66 mol of propene?

Use the mole to mole ratio for C_3H_6 to H_2O given by the equation as a conversion factor to convert from moles of C_3H_6 to moles of H_2O.

$$1.66 \ \cancel{\text{mol } C_3H_6} \ \times \ \frac{1 \ \cancel{\text{mol } H_2O}}{1 \ \cancel{\text{mol } C_3H_6}} \ = \ 1.66 \ \text{mol } H_2O$$

c. How many moles of isopropyl alcohol are expected from the complete reaction of 47.2 g of propene?

Use the molar mass of propene to convert from grams to moles. Then, use the mole to mole ratio for C_3H_6 to C_3H_7OH given by the equation as a conversion factor to convert from moles of C_3H_6 to moles of C_3H_7OH.

$$47.2 \ \cancel{\text{g } C_3H_6} \ \times \ \frac{1 \ \cancel{\text{mol } C_3H_6}}{42.08 \ \cancel{\text{g } C_3H_6}} \ \times \ \frac{1 \ \text{mol } C_3H_7OH}{1 \ \cancel{\text{mol } C_3H_6}} \ = \ 1.12 \ \text{mol } C_3H_7OH$$

d. How many grams of isopropyl alcohol are expected from the complete reaction of 125 g of propene?

Use molar mass of propene to convert from grams to moles. Then, use the mole to mole ratio for C_3H_6 to C_3H_7OH given by the equation as a conversion factor to convert from moles of C_3H_6 to moles of C_3H_7OH. Finally, use the molar mass of 2-propanol to convert moles of 2-propanol to grams.

$$125 \ \cancel{\text{g } C_3H_6} \ \times \ \frac{1 \ \cancel{\text{mol } C_3H_6}}{42.08 \ \cancel{\text{g } C_3H_6}} \ \times \ \frac{1 \ \cancel{\text{mol } C_3H_7OH}}{1 \ \cancel{\text{mol } C_3H_6}} \ \times \ \frac{60.10 \ \text{g } C_3H_7OH}{1 \ \cancel{\text{mol } C_3H_7OH}} \ = \ 179 \ \text{g } C_3H_7OH$$

5.73   *Consider the reaction*

$$KOH(s) \ + \ CO_2(g) \ \rightarrow \ KHCO_3(s)$$

a. How many grams of KOH are required to react completely with 5.00 mol of CO_2?

Use the mole to mole ratio for CO_2 to KOH given by the equation as a conversion factor to convert from moles CO_2 to moles of KOH. Then use the molar mass of KOH to convert moles of KOH to grams.

$$5.00 \; \cancel{\text{mol } CO_2} \; \times \; \frac{1 \; \cancel{\text{mol KOH}}}{1 \; \cancel{\text{mol } CO_2}} \; \times \; \frac{56.11 \text{ g KOH}}{1 \; \cancel{\text{mol KOH}}} \; = \; 281 \text{ g KOH}$$

b. How many grams of $KHCO_3$ are produced from the complete reaction of 75.9 g of KOH?

Use the molar mass of KOH to convert from grams to moles. Then, use the mole to mole ratio for KOH to $KHCO_3$ given by the equation as a conversion factor to convert from moles of KOH to moles of $KHCO_3$. Finally, use the molar mass of $KHCO_3$ to convert moles of $KHCO_3$ to grams.

$$75.9 \; \cancel{\text{g KOH}} \; \times \; \frac{1 \; \cancel{\text{mol KOH}}}{56.11 \; \cancel{\text{g KOH}}} \; \times \; \frac{1 \; \cancel{\text{mol } KHCO_3}}{1 \; \cancel{\text{mol KOH}}} \; \times \; \frac{100.12 \text{ g } KHCO_3}{1 \; \cancel{\text{mol } KHCO_3}} \; = \; 135 \text{ g } KHCO_3$$

5.75   *One form of phosphorus, called white phosphorus, burns when exposed to air.*

$$P_4(s) \; + \; O_2(g) \; \rightarrow \; P_4O_{10}(s)$$

a. Balance the reaction equation.

There are 2 O atoms on the left and 10 O atoms on the right. Place a 5 in front of the O_2 ($5O_2$). This gives ten O atoms on the left. Since the P atoms are already balanced, the equation is balanced.

$$P_4(s) + 5O_2(g) \rightarrow P_4O_{10}(s)$$

b. What is the theoretical yield (in grams) of P_4O_{10} if 33.0 g of P_4 are reacted with 40.0 g of O_2?

The theoretical yield is the amount of product expected when all of the reactant is used up. When it cannot be assumed that another reactant in the reaction is in excess, you must first determine which reactant produces the smaller amount of product when used up. Do this by converting each number of grams given to moles of product formed. Then use the smaller number of moles of product to determine the grams of product that would actually be produced.

$$33.0 \ \cancel{\text{g } P_4} \ \times \ \frac{1 \ \cancel{\text{mol } P_4}}{123.90 \ \cancel{\text{g } P_4}} \ \times \ \frac{1 \ \text{mol } P_4O_{10}}{1 \ \cancel{\text{mol } P_4}} \ = \ 0.266 \ \text{mol } P_4O_{10}$$

$$40.0 \ \cancel{\text{g } O_2} \ \times \ \frac{1 \ \cancel{\text{mol } O_2}}{32.00 \ \cancel{\text{g } O_2}} \ \times \ \frac{1 \ \text{mol } P_4O_{10}}{5 \ \cancel{\text{mol } O_2}} \ = \ 0.250 \ \text{mol } P_4O_{10}$$

Notice that the O_2 will produce the lesser amount of product when it is used up and it will therefore determine the amount of product that can be produced. Convert the 0.250 mol P_4O_{10} to grams using the molar mass of P_4O_{10}.

$$0.250 \ \cancel{\text{mol } P_4O_{10}} \ \times \ \frac{283.89 \ \text{g } P_4O_{10}}{1 \ \cancel{\text{mol } P_4O_{10}}} \ = \ 71.0 \ \text{g } P_4O_{10}$$

5.77   *Magnesium and iodine react to form magnesium oxide.*

$$Mg + I_2 \rightarrow MgI_2$$

a. *What is the limiting reactant if 10.0 g Mg are reacted with 95.0 g of I_2?*

Make sure the equation is balanced. It is in this case. The limiting reactant is the reactant that is completely consumed first and therefore limits the amount of product formed. To determine the limiting reactant in this reaction, first convert the amounts of reactants in grams to moles:

$$10.0 \ \cancel{\text{g Mg}} \ \times \ \frac{1 \ \text{mol Mg}}{24.31 \ \cancel{\text{g Mg}}} \ = \ 0.411 \ \text{mol Mg}$$

$$95.0 \ \cancel{\text{g } I_2} \ \times \ \frac{1 \ \text{mol } I_2}{253.81 \ \cancel{\text{g } I_2}} \ = \ 0.374 \ \text{mol } I_2 \ \leftarrow \text{limiting reactant}$$

Because the balanced equation requires 1 mole of Mg to react with 1 mole of I_2, it is clear that I_2 is the limiting reactant. When 0.374 mol of I_2 has reacted with 0.374 mol Mg, there will be some Mg left unreacted because the original amount (0.411 mol) is greater than 0.374 mol.

b. *What is the theoretical yield of MgI_2 in grams?*

Use the amount of the limiting reactant to calculate the theoretical yield in grams of MgI_2:

$$0.374 \; \cancel{mol \; I_2} \; \times \; \frac{1 \; \cancel{mol \; MgI_2}}{1 \; \cancel{mol \; I_2}} \; \times \; \frac{278.11 \; g \; MgI_2}{1 \; \cancel{mol \; MgI_2}} \; = \; 104 \; g \; MgI_2$$

c. If 25.7 g of MgI_2 are obtained, what is the percent yield?

The percent yield can be calculated using the formula below. The actual yield is 25.7 g and the theoretical yield is 104 g.

$$percent \; yield \; = \; \frac{actual \; yield}{theoretical \; yield} \; \times \; 100\%$$

$$percent \; yield \; = \; \frac{25.7 \; g}{104 \; g} \; \times \; 100\% \; = \; 24.7\%$$

5.79   *If a person drinks a glass of beer or wine, the first step in metabolizing the ethanol (CH_3CH_2OH) is its enzymatic conversion to acetaldehyde (CH_3CHO).*

ethanol   acetaldehyde

a. In this reaction is the ethanol oxidized or reduced?

Oxidized
In the case of an organic molecule, it is not possible to simply look up the charges on the atoms to determine the charges. Oxidation has occurred when H atoms are removed, oxygen is added, or both. Reduction is typically evidenced by the addition of H atoms, removal of oxygen atoms from, or both. In this case H atoms are removed as the molecule is changed from ethanol to acetaldehyde.

b. What theoretical yield of acetaldehyde (in grams) is expected from the reaction of 10.0 g of ethanol? (Assume that ethanol is the limiting reactant.)

Calculate a theoretical yield first by using the molar mass of ethanol to convert grams of ethanol to moles of ethanol. Then convert moles of ethanol to moles of acetaldehyde using the mole to mole ratio from the equation. Finally, convert moles of acetaldehyde to grams of acetaldehyde using the molar mass of acetaldehyde. The molecular formula of ethanol is C_2H_5OH and the molecular formula of acetaldehyde is C_2H_4O.

$$10.0 \text{ g } C_2H_5OH \times \frac{1 \text{ mol } C_2H_5OH}{46.07 \text{ g } C_2H_5OH} \times \frac{1 \text{ mol } C_2H_4O}{1 \text{ mol } C_2H_5OH} \times \frac{44.05 \text{ g } C_2H_4O}{1 \text{ mol } C_2H_4O} = 9.56 \text{ g } C_2H_4O$$

5.81   *Which of the reactions are spontaneous?*

For a spontaneous reaction, ΔG must have a negative value. For a nonspontaneous reaction, ΔG has a positive value.

a. *$CO(g) + H_2O(g) \rightarrow CO_2(g) + H_2(g)$*   *$\Delta G = -4.1 \text{ kcal/mol}$*

This is a spontaneous reaction because the ΔG value is negative.

b. *$2HI(g) \rightarrow H_2(g) + I_2(g)$*   *$\Delta G = 0.6 \text{ kcal/mol}$*

This is a nonspontaneous reaction because the ΔG value is positive.

5.83   *a. Define the term "spontaneous reaction".*

A spontaneous reaction proceeds without outside intervention once started.

b. Define the term "nonspontaneous reaction".

A nonspontaneous reaction requires outside intervention to get it started.

5.85   *Draw an energy diagram for the reaction in Problem 5.81a. Label the x- and y-axis, reactants, products and ΔG.*

The reaction in 5.81a is:

$$CO(g) + H_2O(g) \rightarrow CO_2(g) + H_2(g) \qquad \Delta G = -4.1 \text{ kcal/mol}$$

Energy

CO(g) + H₂O(g)

ΔG = -4.1 kcal/mol

CO₂(g) + H₂(g)

Progress of Reaction

5.87   *Draw two reaction energy diagrams that illustrate thet type of results expected for the reaction below, one diagram for the reaction in the presence of Pt catalyst, and one diagram in the absence of Pt catalyst. In drawing the energy diagrams, assume that the reactions are spontaneous.*

$$CH_3CH=CH_2 + H_2 \xrightarrow{Pt} CH_3CH_2CH_3$$

Reaction energy diagram for the reaction without a catalyst.

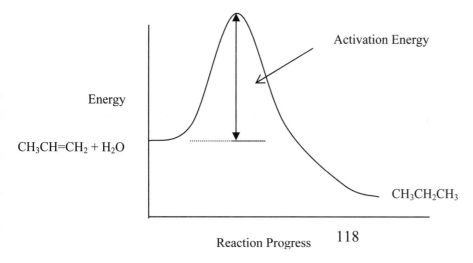

Activation Energy

Energy

CH₃CH=CH₂ + H₂O

CH₃CH₂CH₃

Reaction Progress   118

A catalyst lowers the activation energy for the reaction. In the diagram below, the height of the activation energy is less than that for the diagram of the uncatalyzed reaction.

Reaction energy diagram for the reaction with a catalyst.

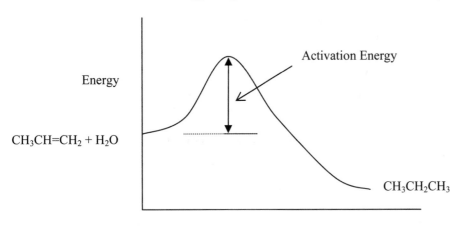

5.89 *What is a catalyst?*

A catalyst is a substance that increases the rate of a reaction without being changed or consumed itself.

5.91 *When hydrogen peroxide is used as a disinfectant, it is the O_2 gas produced from the decomposition of H_2O_2 that kills bacteria. Balance the reaction equation.*
$$H_2O_2(aq) \;\rightarrow\; H_2O(l) \;+\; O_2(g)$$

$2\ H_2O_2(aq) \rightarrow 2\ H_2O(l) \;+\; O_2(g)$

5.93 *Explain the role that carbonic anhydrase has in the transport of CO_2 in the bloodstream.*

Carbonic anhydrase is an enzyme that catalyzes the conversion of CO_2 in red blood cells to H_2CO_3, carbonic acid, which breaks apart into the bicarbonate ion HCO_3^- and H^+. The bicarbonate ion is then released into the blood plasma.

In the lungs, this process is reversed; carbonic anhydrase catalyzes the conversion of H_2CO_3 back to CO_2 which is exhaled along with additional CO_2 dissolved in blood serum and released from its attachment to hemoglobin.

5.95   *The reaction that takes place in a car battery when it starts the car is*

$$Pb + PbO_2 + H_2SO_4 \rightarrow PbSO_4 + H_2O$$

a. Balance the chemical equation.

$Pb + PbO_2 + 2H_2SO_4 \rightarrow 2PbSO_4 + 2\ H_2O$

b. Pb is the atomic symbol for which element?

lead

c. What is the charge on Pb (the first reactant in the reaction equation)?

zero

d. What is the charge on the Pb ion in PbO_2?

+4

e. What is the charge on the Pb ion in $PbSO_4$?

+2

f. In this reaction, what is oxidized?

Pb

g. What is reduced?

Pb^{4+}

h. What is the oxidizing agent?

Pb^{4+}

i. What is the reducing agent?

Pb

j. If 5.0g of Pb are reacted with 5.6 g PbO₂, what is the theoretical yield of PbSO₄? Assume that there is more than enough H₂SO₄ for the reaction to take place.

First, determine the limiting reactant. Convert grams of each reactant to moles.

$$5.0 \ \cancel{g \ Pb} \ \times \ \frac{1 \ mol \ Pb}{207.20 \ \cancel{g \ Pb}} \ = \ 0.024 \ mol \ Pb$$

$$5.6 \ \cancel{g \ PbO_2} \ \times \ \frac{1 \ mol \ PbO_2}{239.20 \ \cancel{g \ PbO_2}} \ = \ 0.023 \ mol \ PbO_2$$

Use the molar ratio from the balanced equation to determine how much PbO_2 is required if all of the Pb is consumed.

$$0.024 \ \cancel{mol \ Pb} \ \times \ \frac{1 \ mol \ PbO_2}{1 \ \cancel{mol \ Pb}} \ = \ 0.024 \ mol \ PbO_2$$

Because the amount of PbO_2 required is greater than what is present, PbO_2 is the limiting reactant. Use the amount of PbO_2 to calculate the theoretical yield of $PbSO_4$.

$$0.023 \ \cancel{mol \ PbO_2} \ \times \ \frac{2 \ \cancel{mol \ PbSO_4}}{1 \ \cancel{mol \ PbO_2}} \ \times \ \frac{303.26 \ g \ PbSO_4}{1 \ \cancel{mol \ PbSO_4}} \ = \ 14 \ g \ PbSO_4$$

k. If 4.5 g of PbSO₄ are obtained in the reaction described in part j, what is the percent yield?

$$percent \ yield \ = \ \frac{4.5 \ g}{14 \ g} \ \times \ 100\% \ = \ 32\%$$

l. Would you expect ΔG for this reaction to have a negative or positive value?

Negative. The reaction is spontaneous which is the basis for its use in batteries.

m. When it is in use, a battery heats up. Is the reaction exothermic or endothermic?

Exothermic. Heat is released as the reaction occurs.

n. A running car engine recharges the battery by reversing the reaction above. Would you expect ΔG for this process to have a negative or positive value?

Positive. The reverse reaction is nonspontaneous as it requires an external source of driving force (car engine) to occur.

5.97 *As we will see in a later chapter, high temperatures can cause a protein to lose its normal biological function. This being the case, explain why an increase in temperature does not always lead to an increased reaction rate for biochemical reactions.*

An increase in temperature can cause a protein to lose its normal biological function. Once this happens, any reaction that the protein participates in is affected and the reaction rate can fail to increase or it can even slow down.

5.99 *In dilute nitric acid, HNO_3, copper metal dissolves according to the following equation:*

$$3\ Cu(s)\ +\ 8\ HNO_3(aq)\ \rightarrow\ 3\ Cu(NO_3)_2(aq)\ +\ 2\ NO(g)\ +\ 4\ H_2O(l)$$

How many grams of HNO_3 are needed to dissolve 11.45 g of Cu according to this equation?

First, calculate the moles of Cu that need to be dissolved. Then, calculate the moles of HNO_3 required to dissolve all of the Cu using the molar ratio from the balanced equation. Convert the moles of HNO_3 to grams using the molar mass of HNO_3.

$$11.45\ \text{g Cu}\ \times\ \frac{1\ \text{mol Cu}}{63.55\ \text{g Cu}}\ \times\ \frac{8\ \text{mol } HNO_3}{3\ \text{mol Cu}}\ \times\ \frac{63.01\ \text{g}}{1\ \text{mol } HNO_3}\ =\ 30.28\ \text{g } HNO_3$$

Chapter 6
Gases, Solutions, Colloids, and Suspensions

Solutions to Problems

6.1   *a. For the gas shown below, has the volume increased, decreased, or remained the same?*

Decreased as indicated by the lower position of the piston.

b. Has the pressure inside the container increased, decreased, or remained the same?

Increased. The particles occupy a smaller volume and will collide with the inner walls with higher frequency (as long as the temperature remains the same).

c. Have the number of moles increased, decreased, or remained the same?

Remained the same. The number of moles of gas particles should stay the same as long as it can be assumed that the container is tightly sealed.

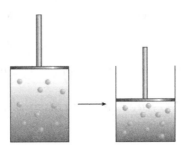

6.3   *a. What is heat of fusion?*

Heat of fusion is the energy required to melt a solid.

b. What is heat of vaporization?

Heat of vaporization is the energy required to evaporate a liquid.

6.5   *If you immerse your arm in a bucket of ice water, your arm gets cold. Where does the heat energy from your arm go and what process is the energy used for?*

The heat energy goes into the ice and the energy is used in the melting process.

6.7   *True or False? If heat is continually added to a pan of boiling water, the temperature of the water continually rises until all of the water has boiled away.*

False. Once the water starts to boil, the temperature remains constant. At the boiling point, the heat energy goes into separating the molecules as the state changes from liquid to gas.

6.9   *At room temperature pentane ($CH_3CH_2CH_2CH_2CH_3$) is a liquid, while methane (CH_4) is a gas.*

a.   *Which alkane has a higher boiling point?*

Pentane ($CH_3CH_2CH_2CH_2CH_3$). Pentane is a liquid at room temperature which means that its boiling point is higher than room temperature. Methane is a gas at room temperature which means that its boiling point is lower than room temperature.

b.   *When methane boils, what is being broken: a covalent bond, a polar covalent bond, an ionic bond, or noncovalent interaction?*

Noncovalent interaction. Boiling is a physical process which does not involve a change in chemical identity. Breaking covalent or ionic bonds results in a new substance.

c.   *Name the force that holds one pentane molecule to another.*

London force. This is the only intermolecular force present between nonpolar molecules.

d.   *Account for the difference in the boiling points of pentane and methane.*

Pentane has a higher boiling point because the strength of the London force between the larger pentane molecules is greater.

6.11   *Define STP.*

STP or standard temperature and pressure is defined to be 1 atm (or 14.7 psi or 760 torr) and 0°C.

6.13   *A pressure of 9.2 psi is how many*

Use the following information to set up the conversion calculations:

1 atm = 760 torr or mmHg = 14.7 psi

a. atmospheres?

$$9.2 \; \cancel{psi} \times \frac{1 \text{ atm}}{14.7 \; \cancel{psi}} = 0.63 \text{ atm}$$

b. torr?

$$9.2 \; \cancel{psi} \times \frac{760 \text{ torr}}{14.7 \; \cancel{psi}} = 480 \text{ torr}$$

6.15   *A pressure of 814 torr is how many*

Use the following information to set up the conversion calculations:

1 atm = 760 torr or mmHg = 14.7 psi

a.   atmospheres?

$$814 \; \cancel{torr} \times \frac{1 \text{ atm}}{760 \; \cancel{torr}} = 1.07 \text{ atm}$$

b.   psi?

$$814 \; \cancel{torr} \times \frac{14.7 \text{ psi}}{760 \; \cancel{torr}} = 15.7 \text{ psi}$$

6.17   *In the SI measurement system, the unit of pressure is called the kilopascal (kPa). One atmosphere equals 101.3 kPa.*

a.   Convert 1.29 atm into kilopascals.

$$1.29 \; \cancel{atm} \times \frac{101.3 \text{ kPa}}{1 \; \cancel{atm}} = 131 \text{ kPa}$$

b. *Convert 87.2 kPa into atmospheres.*

$$87.2 \ \cancel{kPa} \quad \times \quad \frac{1 \ atm}{101.3 \ \cancel{kPa}} \quad = \quad 0.861 \ atm$$

c. *Convert 612 torr into kilopascals.*

$$612 \ \cancel{torr} \quad \times \quad \frac{1 \ \cancel{atm}}{760 \ \cancel{torr}} \quad \times \quad \frac{101.3 \ kPa}{1 \ \cancel{atm}} \quad = \quad 81.6 \ kPa$$

6.19   *Explain how a barometer works.*

A barometer is an instrument that measures atmospheric pressure. It is constructed by inverting a glass tube containing mercury into a dish that also contains mercury. The height of the mercury inside the tube will go up and down depending on how much atmospheric pressure is pushing down on the mercury in the dish. When the atmospheric pressure is equal to 1 atm, the height of the mercury in the tube will be 760 mm.

6.21   *At an atmospheric pressure of 760 torr and a temperature of 0°C, the mercury level in the right arm of the manometer pictured in Figure 6.7b is 5 mm higher than the mercury in the left arm. What is the gas pressure inside the flask?*

760 torr + 5 torr = 765 torr

The first thing to note is that the manometer pictured in Figure 6.7b is an open-ended manometer. This means that the gas pressure in the gas bulb is pushing against the outside atmospheric pressure of 760 torr. Since the mercury moved up on the right arm this tells you that the pressure of the gas in the bulb is 5 mmHg greater than atmospheric pressure and this much pressure can be added to the atmospheric pressure to get the pressure of the gas. The second thing to note is that the atmospheric pressure is in torr, but the torr and mmHg are interchangeable units.

6.23 *Use Table 6.1 to estimate the boiling point of water at a pressure of 920 torr.*

105°C
The boiling point of a liquid is the temperature at which the vapor pressure equals the pressure of the atmosphere above it. Table 6.1 gives the vapor pressure of water at various temperatures. When the pressure of the atmosphere above it is 920 torr, water will boil at a temperature of about 105°C.

6.25 *If you go camping in the mountains, why does it take longer to cook a pot of noodles than it does at sea level?*

At higher altitudes, the atmospheric pressure is lower than at sea level. When cooking noodles in boiling water at higher altitudes, the water will boil at a lower temperature. In order to provide sufficient heat for the noodles to cook (equal to the amount of heat it would take at sea level), the process will require more time.

6.27 *At a pressure of 760 torr a balloon has a volume of 1.50 L. If the balloon is put into a container and the pressure is increased to 2500 torr (at constant temperature), what is the new volume of the balloon?*

Assign values to the variables.
$P_1 = 760$ torr $P_2 = 2500$ torr
$V_1 = 1.50$ L $V_2 = ?$ (this is what you are asked to calculate)

Select the gas law that has these variables ONLY. In this case it will be **Boyle's Law; $P_1V_1 = P_2V_2$** . Rearrange the equation so that the required variable is by itself. Replace the variables in the equation with the values you identified above and solve for the one missing.

$$V_2 = \frac{P_1 V_1}{P_2}$$

$$V_2 = \frac{760 \text{ torr} \times 1.50 \text{ L}}{2500 \text{ torr}} = 0.46 \text{ L}$$

6.29   *At a temperature of 30°C, a balloon has a volume of 1.50 L. If the temperature is increased to 60°C (at constant pressure), what is the new volume of the balloon?*

In gas law problems the temperature must be converted to Kelvin.
K = °C + 273.15 = 30°C + 273 = 303 K
K = °C + 273.15 = 60°C + 273 = 333 K

Assign values to the variables.
$T_1 = 303$ K $T_2 = 333$ K
$V_1 = 1.50$ L $V_2 = ?$ (this is what you are asked to calculate)

Select the gas law that has these variables ONLY. In this case it will be **Charles' Law; $V_1 / T_1 = V_2 / T_2$**. Rearrange the equation so that the required variable is by itself. Replace the variables in the equation with the values you identified above and solve for the one missing.

$$V_2 = \frac{V_1 T_2}{T_1}$$

$$V_2 = \frac{333 \text{ K} \times 1.50 \text{ L}}{303 \text{ K}} = 1.65 \text{ L}$$

6.31   *At a temperature of 30°C, a gas inside a 1.50 L metal canister has a pressure of 760 torr. If the temperature is increased to 60°C (at constant volume), what is the new pressure of the gas?*

In gas law problems the temperature must be converted to kelvins.
K = °C + 273.15 = 30°C + 273 = 303 K
K = °C + 273.15 = 60°C + 273 = 333 K

Assign values to the variables.
$T_1 = 303$ K $T_2 = 333$ K
$P_1 = 760$ torr $P_2 = ?$ (this is what you are asked to calculate)

Note: even though the volume was stated it is not included as a variable because the problem states that it remained constant.

Select the gas law that has these variables ONLY. In this case it will be **Gay-Lussac's Law; $P_1/T_1 = P_2/T_2$**. Rearrange the equation so that the required variable is by itself. Replace the variables in the equation with the values you identified above and solve for the one missing.

$$P_2 = \frac{P_1 T_2}{T_1}$$

$$V_2 = \frac{760 \text{ torr} \times 333 \text{ K}}{303 \text{ K}} = 835 \text{ torr}$$

6.33   *A 2.0 L balloon contains 0.35 mol of $Cl_2(g)$. At constant pressure and temperature, what is the new volume of the balloon if 0.20 mol of gas is removed?*

Identify the variables.
Note that 0.20 mol is the amount removed, NOT final number of moles.

$n_2 = 0.35 - 0.20 = 0.15$ mol

$\quad\quad\quad$ $n_1 = 0.35$ mol $\quad\quad\quad$ $n_2 = 0.15$ mol
$\quad\quad\quad$ $V_1 = 2.0$ L $\quad\quad\quad\quad$ $V_2 = ?$ (this is what you are asked to calculate)

Select the gas law that has these variables ONLY. In this case it will be **Avogadro's Law; $V_1/n_1 = V_2/n_2$**. Rearrange the equation so that the required variable is by itself. Replace the variables in the equation with the values you identified above and solve for the one missing.

$$V_2 = \frac{V_1 n_2}{n_1}$$

$$V_2 = \frac{2.0 \text{ L} \times 0.15 \text{ mol}}{0.35 \text{ mol}} = 0.86 \text{ L}$$

6.35　*A balloon with a volume of 1.50 L is at a pressure of 760 torr and a temperature of 30°C. If the balloon is put into a container and the pressure is increased to 2500 torr and the temperature is raised to 60 °C, what is the new volume of the balloon?*

First convert °C to K.
K = °C + 273.15 = 30°C + 273 = 303 K
K = °C + 273.15 = 60°C + 273 = 333 K

Assign values to the variables.

T_1 = 303 K	T_2 = 333 K
P_1 = 760 torr	P_2 = 2500
V_1 = 1.50 L	V_2 = ? (this is what you are asked to calculate)

Select the gas law that has all these variables. In this case it will be the **Combined Gas Law; $P_1V_1 / n_1T_1 = P_2V_2 / n_2T_2$**. Rearrange the equation so that the required variable is by itself. Replace the variables in the equation with the values you identified above and solve for the one missing. Note that it is not necessary to include n_1 and n_2 since the number of moles is constant.

$$V_2 = \frac{P_1 V_1 T_2}{T_1 P_2}$$

$$V_2 = \frac{760 \text{ torr} \times 1.50 \text{ L} \times 333 \text{ K}}{2500 \text{ torr} \times 303 \text{ K}} = 0.50 \text{ L}$$

6.37　*A 575 mL metal can contains 2.50×10^{-2} mol of He at a temperature of 298K. What is the pressure (in atm) inside the can? Is this pressure greater than or less than standard atmospheric pressure?*

Assign values to the variables.

T = 298 K	P = ?
V = 0.575 L	n = 2.50×10^{-2} mol

Select the gas law that has just one of each variable. In this case it will be the **Ideal Gas Law; PV = n R T**. R is the gas constant (0.0821 L atm /mol K). Rearrange the equation so that the required variable is by itself. Replace the variables in the equation with the values you identified above and solve for the one missing.

$$P = \frac{nRT}{V}$$

Note: since the R constant has the units atm and L in it, before solving for P the volume unit must be converted to L.

$$575 \text{ mL} \times \frac{1 \times 10^{-3} \text{ L}}{1 \text{ mL}} = 0.575 \text{ L}$$

$$P = \frac{2.50 \times 10^{-2} \text{ mol} \times (0.0821 \text{ L atm/mol K}) \times 298 \text{ K}}{0.575 \text{ L}} = 1.06 \text{ atm}$$

1.06 atm is greater than standard atmospheric pressure.

6.39   *A 250.0 mL flask contains 0.350 mol of O_2 at 40°C.*

Use the ideal gas equation to calculate the pressure, which is the only unknown property of the gas. Use the conversion equivalences: 1 atm = 760 torr = 14.7 psi to convert the pressure from one unit to another.

a. What is the pressure in atm?

$PV = nRT$

$P = ?$ $V = 250.0 \text{ mL} = 2.500 \times 10^{-1} \text{ L}$ $n = 0.350 \text{ mol } O_2$

$T = 40°C + 273.15 = 310 \text{ K}$

$PV = nRT$ therefore $P = \dfrac{nRT}{V}$

$$P = \frac{0.350 \text{ mol} \times 0.0821 \text{ L} \cdot \text{atm/mol} \cdot \text{K} \times 310 \text{ K}}{2.500 \times 10^{-1} \text{ L}} = 36 \text{ atm}$$

b. What is the pressure in torr?

1 atm = 760 torr

$$36 \text{ atm} \times \frac{760 \text{ torr}}{1 \text{ atm}} = 2.7 \times 10^{4} \text{ torr}$$

131

c. What is the pressure in psi?

1 atm = 14.7 psi

$$36 \; \text{atm} \;\times\; \frac{14.7 \; \text{psi}}{1 \; \text{atm}} \;=\; 530 \, \text{psi}$$

6.41 *A 250 mL flask contains He at a pressure of 760 torr and a temperature of 25ºC. What mass of He is present?*

First, use the ideal gas equation to calculate the moles of He present. Then, use the molar mass of He to convert to grams of He.

$P = 760 \text{ torr} = 1.0 \text{ atm}, \; V = 250 \text{ mL} = 0.25 \text{ L}, \; T = 25°C + 273.15 = 298 \text{ K}$

$$PV = nRT \quad \text{therefore} \quad n \;=\; \frac{PV}{RT} \;=\; \frac{1.0 \text{ atm x } 0.25 \text{ L}}{0.0821 \text{ L} \cdot \text{atm/mol} \cdot \text{K x } 298 \text{ K}} \;=\; 0.010 \, \text{mol He}$$

$$0.010 \; \text{mol} \; \text{He} \;\times\; \frac{4.003 \text{ g}}{1 \; \text{mol}} \;=\; 0.040 \, \text{g He}$$

6.43 *a. How many moles of Ar are present in a 2.0 L flask that has a pressure of 1.05 atm at a temperature of 25°C?*

$P = 1.05 \text{ atm}, \; V = 2.0 \text{ L}, \; T = 25°C + 273.15 = 298 \text{ K}$

$$PV = nRT \quad \text{therefore} \quad n \;=\; \frac{PV}{RT}$$

$$n \;=\; \frac{1.05 \text{ atm x } 2.0 \text{ L}}{0.0821 \text{ L} \cdot \text{atm/mol} \cdot \text{K x } 298 \text{ K}} \;=\; 0.086 \, \text{mol Ar}$$

b. What is the mass of this Ar?

$$0.086 \; \text{mol} \; \text{Ar} \;\times\; \frac{39.95 \text{ g}}{1 \; \text{mol}} \;=\; 3.4 \, \text{g Ar}$$

6.45 *A sample of a gas at a temperature of 25.0°C has a pressure of 815 torr and occupies a volume of 9.92 L.*

a. Use Boyle's law to calculate the new pressure if the temperature is held constant and the volume is decreased to 5.92 L.

Boyle's law: $\qquad$ $P_1V_1 = P_2V_2$
In this problem: $\qquad$ $P_1 = 815$ torr $\qquad\qquad$ $P_2 = ?$
$\qquad\qquad\qquad\qquad$ $V_1 = 9.92$ L $\qquad\qquad$ $V_2 = 5.92$ L

Solving for P_2:

$$P_2 = \frac{P_1V_1}{V_2} = \frac{815 \text{ torr} \times 9.92\text{ }\cancel{L}}{5.92\text{ }\cancel{L}} = 1.37 \times 10^3 \text{ torr}$$

b. Use Gay-Lussac's law to calculate the new pressure if the volume is held constant and the temperature is increased to 125.0°C.

Gay-Lussac's law: $\qquad \dfrac{P_1}{T_1} = \dfrac{P_2}{T_2}$

In this problem:
$P_1 = 815$ torr $\qquad\qquad\qquad\qquad$ $P_2 = ?$
$T_1 = 25.0°C + 273.15 = 298.2$ K $\qquad$ $T_2 = 125.0°C + 273.15 = 398.2$ K

Solving for P_2:

$$P_2 = \frac{P_1}{T_1} \times T_2 = \frac{815 \text{ torr}}{298.2 \text{ K}} \times 398.2 \text{ K} = 1.09 \times 10^3 \text{ torr}$$

c. Use Charles' law to calculate the new volume if the pressure is held constant and the temperature is increased to 125.0°C.

Charles' law: $\qquad \dfrac{V_1}{T_1} = \dfrac{V_2}{T_2}$

In this problem:
$V_1 = 9.92$ L $\qquad\qquad\qquad\qquad$ $V_2 = ?$
$T_1 = 25.0°C + 273.15 = 298.2$ K $\qquad$ $T_2 = 125.0°C + 273.15 = 398.2$ K

Solving for P_2:

$$V_2 = \frac{V_1}{T_1} \times T_2 = \frac{9.92 \text{ L}}{298.2 \text{ K}} \times 398.2 \text{ K} = 13.2 \text{ L}$$

d. *Use the combined gas law to calculate the new pressure if the temperature is increased to 125.0°C and the volume is decreased to 5.92 L.*

Combined gas law: $\dfrac{P_1 V_1}{T_1} = \dfrac{P_2 V_2}{T_2}$

In this problem:

$P_1 = 815$ torr	$P_2 = ?$
$V_1 = 9.92$ L	$V_2 = 5.92$ L
$T_1 = 25.0°C + 273.15 = 298.2$ K	$T_2 = 125.0°C + 273.15 = 398.2$ K

Solving for P_2:

$$P_2 = \frac{P_1 V_1}{T_1} \ \text{x}\ \frac{T_2}{V_2} = \frac{815 \text{ torr x } 9.92 \text{ L}}{298.2 \text{ K}} \ \text{x}\ \frac{398.2 \text{ K}}{5.92 \text{ L}} = 1.82 \text{ x } 10^3 \text{ torr}$$

e. *Use the ideal gas law to calculate the number of moles of gas that are present.*

Ideal gas law: $PV = nRT$

First, we have to convert the pressure from torr to atm:

$$815 \text{ torr} \ \text{x}\ \frac{1 \text{ atm}}{760 \text{ torr}} = 1.07 \text{ atm}$$

Now, using the ideal gas law to solve for the number of moles, n:

$$n = \frac{PV}{RT} = \frac{1.07 \text{ atm x } 9.92 \text{ L}}{(0.0821 \text{ L atm /mol K) x } 298.2 \text{ K}} = 0.434 \text{ mol}$$

6.47 *The cover of the rock band Led Zeppelin's first album pictured the Hindenburg, a hydrogen-filled dirigible, going down in flames. At the temperature of 25°C and a pressure of 760 torr, the Hindenburg held 7.062 x 10^6 cubic feet of hydrogen gas. How many grams of hydrogen did this represent?*

Use the ideal gas law to calculate the moles of hydrogen gas (H_2) using the given information. Then, use the molar mass of H_2 to calculate the mass of the hydrogen gas in the dirigible.

First, convert the quantities given to units appropriate for the ideal gas law equation.

$T = 25°C + 273 = 298$ K

$$P = 760 \text{ torr} \ \ \text{x} \ \ \frac{1 \text{ atm}}{760 \text{ torr}} \ \ = \ \ 1.0 \text{ atm}$$

$$V = 7.062 \text{ x } 10^6 \text{ ft}^3 \ \text{x} \ \left(\frac{12 \text{ in}}{1 \text{ ft}}\right)^3 \ \text{x} \ \left(\frac{2.54 \text{ cm}}{1 \text{ in}}\right)^3 \ \text{x} \ \frac{1 \text{ mL}}{1 \text{ cm}^3} \ \text{x} \ \frac{10^{-3} \text{ L}}{1 \text{ mL}} \ = \ 2.000 \text{ x } 10^8 \text{ I}$$

Using the ideal gas law, calculate the moles of H_2.

$$n \ = \ \frac{PV}{RT} \ = \ \frac{1.0 \text{ atm x } 2.000 \text{ x } 10^8 \text{ L}}{(0.0821 \text{ L atm /mol K}) \text{ x } 298 \text{ K}} \ = \ 8.2 \text{ x } 10^6 \text{ mol}$$

Calculate the grams of H_2 gas using the molar mass of H_2.

$$8.2 \text{ x } 10^6 \text{ mol} \ \text{x} \ \frac{2.0 \text{ g}}{1 \text{ mol}} \ = \ 1.6 \text{ x } 10^7 \text{ g } H_2 \text{ gas}$$

6.49   *True or false? Ten liters of helium gas at STP has the same mass as ten liters of neon gas at STP. Explain.*

False. The same number of moles of each gas is present, but the molar mass of each is different.

6.51   *Use concepts discussed in this chapter to explain why an empty plastic milk bottle collapses if the air is pumped out of it.*

The air around us exerts pressure on everything with which it is in contact, including an empty plastic milk bottle. When open to the atmosphere around it, the pressure of the air inside the bottle exerts a pressure on the inside walls of the empty bottle equal to the pressure exerted by the atmosphere on the outside of the bottle. However, when air inside is removed, that balance is lost. The pressure on the outside of the empty bottle pushes the bottle inward and bottle collapses.

6.53   *A mixture of gases contains 0.75 mol of N_2, 0.25 mol of O_2, and 0.25 mol of He.*

a. What is the partial pressure of each gas (in atm and in torr) in a 25 L cylinder at 350 K?

The partial pressure of a gas can be calculated as though it is the only gas in the container.

Assign values to the variables. For calculating the partial pressure of N_2:

T = 350 K   P = ?
V = 25 L   n = 0.75 mol

Select the gas law that has just one of each variable. In this case it will be the **Ideal Gas Law; PV = n R T**. R is the gas constant (0.0821 L atm /mol K). Rearrange the equation so that the required variable is by itself. Replace the variables in the equation with the values you identified above and solve for the one missing.

$$P_{N2} = \frac{nRT}{V} = \frac{0.75 \; \text{mol} \; \times \; 0.0821 \; \text{L} \cdot atm/\text{mol} \cdot \text{K} \; \times \; 350 \; \text{K}}{25 \; \text{L}} = 0.86 \, atm$$

P_{N2} = 0.86 atm for N_2

Convert to torr: $P_{N2} = 0.86 \; \text{atm} \quad \times \quad \dfrac{760 \, torr}{1 \; \text{atm}} = 6.5 \times 10^2 \, torr$

$P_{N2} = 6.5 \times 10^2$ torr for N_2

Now repeat the calculation for O_2. Since the temperature and volume are the same, you can use the same set-up from above and simply insert the correct number of moles for O_2.

$$P_{O2} = \frac{nRT}{V} = \frac{0.25 \; \text{mol} \; \times \; 0.0821 \; \text{L} \cdot atm/\text{mol} \cdot \text{K} \; \times \; 350 \; \text{K}}{25 \; \text{L}} = 0.29 \, atm$$

P_{O2} = 0.29 atm for O_2

Convert to torr: $P_{O2} = 0.29 \; \text{atm} \quad \times \quad \dfrac{760 \, torr}{1 \; \text{atm}} = 2.2 \times 10^2 \, torr$

$P_{O2} = 2.2 \times 10^2$ torr for O_2

Now repeat the calculation for He. Since the temperature and volume are the same, you can use the same set-up from above and simply insert the correct number of moles for He.

$$P_{He} = \frac{nRT}{V} = \frac{0.25 \; \text{mol} \; \times \; 0.0821 \; \text{L} \cdot atm/\text{mol} \cdot \text{K} \; \times \; 350 \; \text{K}}{25 \; \text{L}} = 0.29 \, atm$$

P_{He} = 0.29 atm for He

Convert to torr: P_{He} = 0.29 ~~atm~~   x $\dfrac{760 \text{ torr}}{1 \text{ ~~atm~~}}$ =   2.2 x 10² torr

P_{He} = 2.2 x 10² torr for He

b. *What is the total pressure?*

Dalton's Law states that the total pressure for all gases in a container is equal to the sum of their partial pressures. Therefore the total pressure is found by adding up the partial pressures calculated above.

$$P_{total} \;=\; P_{N_2} \;+\; P_{O_2} \;+\; P_{He}$$

P_{total} = 0.86 atm + 0.29 atm + 0.29 atm = 1.44 atm

1.44 atm; 1.09 x 10³ torr

6.55   *For the mixture of gases in Problem 6.53 what are the partial pressures and the total pressure if 0.50 mol of CO₂(g) is added?*

First, since gas pressures are independent of each other, the partial pressures of the first three do not change. Calculate the partial pressure of CO₂. Then add this pressure to the partial pressures of the other gases to find the total pressure.

$$P_{CO2} \;=\; \frac{0.50 \text{ ~~mol~~} \times 0.0821 \text{ ~~L~~} \cdot \text{atm}/\text{~~mol~~} \cdot \text{~~K~~} \times 350 \text{ ~~K~~}}{25 \text{ ~~L~~}} \;=\; 0.58 \text{ atm}$$

P_{N2} = 0.86 atm N₂, 6.5 x 10² torr
P_{O2} = 0.29 atm O₂, 2.2 x 10² torr
P_{He} = 0.29 atm He, 2.2 x 10² torr
P_{CO2} = 0.58 atm CO₂, 4.4 x 10² torr

Total Pressure = 2.02 atm = 1.54 x 10³ torr

6.57   *Scuba divers sometimes breathe a gas mixture called trimix, which consists of helium, oxygen, and nitrogen gases. If a scuba tank with a volume of 2.5 L, a pressure of 2700 psi, and a temperature of 45°F contains 27% O_2, 12% He, and 61% N_2,*

a.   *what is the partial pressure of each gas?*

The partial pressure of each gas can be determined using the percentage of the total pressure given for each gas. First, convert psi to atm for further calculations:

$$2700 \text{ psi} \quad \times \quad \frac{1 \text{ atm}}{14.7 \text{ psi}} \quad = \quad 184 \text{ atm}$$

The partial pressure of each gas can then be calculated using their respective percentages:

P_{O_2} = 0.27 x 184 atm = 50. atm

P_{He} = 0.12 x 184 atm = 22 atm

P_{N_2} = 0.61 x 184 atm = 110 atm

b.   *how many grams of each gas are present?*

To determine the grams of each gas, calculate the number of moles then convert from moles to grams using the molar mass of each gas. First, convert the temperature from °F to K:

$$\frac{45°F - 32}{1.8} = 7.2°C \qquad 7.2°C + 273.15 \text{ K} = 280.4 \text{ K}$$

Then, use the ideal gas law to calculate the number of moles given the other know values:

$$n_{O_2} \quad = \quad \frac{P_{O_2} V}{RT} \quad = \quad \frac{50. \text{ atm x } 2.5 \text{ L}}{(0.0821 \text{ L atm /mol K}) \text{ x } 280.4 \text{ K}} \quad = \quad 5.4 \text{ mol } O_2$$

Similar calculations for He and N_2 yield:

n_{He} = 2.4 mol He
n_{N2} = 12 mol N_2

Convert moles to grams by using their respective molar masses:

$$\text{grams of } O_2 = 5.4 \text{ mol} \times \frac{32.0 \text{ g}}{1 \text{ mol}} = 170 \text{ g } O_2$$

$$\text{grams of He} = 2.4 \text{ mol} \times \frac{4.0 \text{ g}}{1 \text{ mol}} = 9.6 \text{ g He}$$

$$\text{grams of } N_2 = 12 \text{ mol} \times \frac{28.0 \text{ g}}{1 \text{ mol}} = 340 \text{ g } N_2$$

c. *how many molecules of 0 and N and atoms of He are present?*

$$\text{molecules of } O_2 = 5.4 \text{ mol} \times \frac{6.02 \times 10^{23} \text{ molecules}}{1 \text{ mol}} = 3.2 \times 10^{24} \text{ molecules } O_2$$

$$\text{atoms of He} = 2.4 \text{ mol} \times \frac{6.02 \times 10^{23} \text{ atoms}}{1 \text{ mol}} = 1.4 \times 10^{24} \text{ molecules He}$$

$$\text{molecules of } N_2 = 12 \text{ mol} \times \frac{6.02 \times 10^{23} \text{ molecules}}{1 \text{ mol}} = 7.2 \times 10^{24} \text{ molecules } N_2$$

6.59   *Define the terms solute and solvent.*

The solute is the component dissolved in the solution. The solvent is the component in the greatest amount in the solution.

6.61   *Define the terms pure substance and mixture.*

A pure substance is made up of just one element or compound. A mixture is a combination of two or more pure substances.

6.63   *You add sugar to a pan of cold water and obtain a saturated solution. When you heat the pan on the stove, all of the sugar dissolves. Explain why this happens.*

The solubility of a solute particle in a given solvent depends on temperature. For gas solutes, the higher the temperature is, the lower their solubility is. For liquid and solid solutes, increasing the temperature increases the solubility.

The solubility of sugar, a solid solute, in water is lower at room temperature than at a higher temperature. Thus, when a saturated solution of sugar is heated, the increase in temperature results in higher solubility and all of the sugar dissolves.

6.65   *Give an example of a solution in which*

   a. *the solute is a solid and the solvent is a liquid.*

   Examples will vary. Some possible answers are: Salt in water, sugar in tea, sugar in coffee.

   b. *the solute is a gas and the solvent is a gas.*

   Examples will vary. Some possible answers are: CO_2 in air, O_2 in air, He in N_2 in a scuba tank.

6.67   *At a pressure of 1 atm, in which is O_2 gas more soluble, 0^oC water 90^oC water?*

   0 °C water. The solubility of a gas is higher when the solution is at a lower temperature.

6.69   a. *If 45.0 g of NaCl are added to 100g of water at 20^oC, will all of the NaCl dissolve?*

   No. Only 35.9 g of NaCl dissolves per 100 g of water at this temperature.

   b. *Is the resulting mixture homogeneous or heterogeneous?*

   Heterogeneous because some of the NaCl remains undissolved in the solid form.

6.71   *Predict whether each ionic compound is soluble or insoluble in water.*

   Use the general solubility rules for ions given in the textbook.

   a. *$(NH_4)_2SO_4$*   Soluble

   b. *K_2SO_4*   Soluble

c. CaCO₃   Insoluble

d. NaNO₃   Soluble

6.73   *When aqueous copper(II) sulfate is mixed with aqueous sodium sulfide, a*
        precipitate forms. Write a balanced equation for this reaction.

Before attempting to balance the equation make sure that you write the correct
formulas. Copper(II) sulfate is $CuSO_4$ (equal numbers of Cu^{2+} and SO_4^{2-} to give a
neutral compound). Sodium sulfide is Na_2S (twice as many Na^+ as S^{2-} to give a
neutral compound). Now that you have the reactants identified you need to predict
the products. A double replacement reaction can occur when two aqueous
solutions of ionic compounds are mixed together. In this case Cu^{2+} combines
with S^{2-} to form CuS and Na^+ combines with the SO_4^{2-} ion to form Na_2SO_4. Since
you are told that a precipitate forms and Table 7.1 shows Na_2SO_4 to be soluble,
the CuS must be the solid formed. Now we can write the equation.

$$CuSO_4(aq) + Na_2S(aq) \rightarrow CuS(s) + Na_2SO_4(aq)$$

We must then determine if the equation is balanced. Start by counting each atom.

reactant atoms	product atoms
1 Cu atom	1 Cu atom
2 S atoms	2 S atoms
4 O atoms	4 O atoms
2 Na atoms	2 Na atoms

In this case, all of the atoms are balanced on both sides of the equation. The
equation is correct as written.

$$\mathbf{CuSO_4(aq) + Na_2S(aq) \rightarrow CuS(s) + Na_2SO_4(aq)}$$

6.75   *Complete and balance each precipitation reaction.*

All of the reactions involve mixing two aqueous solutions of ionic compounds. In
each case a double replacement reaction can occur. Follow the same basic

procedure for each of the equations below. Recombine the reactant cations and anions to form new ionic compounds. Refer to Table 7.1 to see if either compound formed is insoluble. If it is insoluble be sure to write *(s)* with the formula. Use the symbol *(aq)* for those that are water soluble. Then, write the balanced equation.

a. $CaCl_2(aq) + Li_2CO_3(aq) \rightarrow$

Ca^{2+} reacts with CO_3^{2-} to form $CaCO_3$ and Li^+ reacts with Cl^- to form $LiCl$. The solubility Table 7.1 indicates that calcium carbonate is insoluble but lithium chloride is soluble.

$CaCl_2(aq) + Li_2CO_3(aq) \rightarrow CaCO_3(s) + LiCl(aq)$

Now balance the reaction.
> *(NOTE: do not start balancing until you have written the formulas of all reactants and products.)*

$CaCl_2(aq) + Li_2CO_3(aq) \rightarrow CaCO_3(s) + LiCl(aq)$

Start by counting each atom.

reactant atoms	product atoms
1 Ca atom	1 Ca atom
2 Cl atoms	1 Cl atoms
3 O atoms	3 O atoms
2 Li atoms	1 Li atoms
1 C atom	1 C atom

There are 2 Cl atoms on the left and 1 on the right, so multiply LiCl by 2 (2LiCl) and recount.

reactant atoms	product atoms
1 Ca atom	1 Ca atom
2 Cl atoms	**2 Cl atoms**
3 O atoms	3 O atoms
2 Li atoms	**2 Li atoms**
1 C atom	1 C atom

This makes the atom count the same on both sides, so the equation is balanced.

$CaCl_2(aq) + Li_2CO_3(aq) \rightarrow CaCO_3(s) + 2LiCl(aq)$

b. $Pb(NO_3)_2(aq) + NaCl(aq) \rightarrow$

Pb^{2+} reacts with Cl^- to form $PbCl_2$ and Na^+ reacts with NO_3^- to form $NaNO_3$. The solubility Table 7.1 indicates that $PbCl_2$ is insoluble and that the other compounds are soluble.

$Pb(NO_3)_2(aq) + NaCl(aq) \rightarrow PbCl_2(s) + NaNO_3(aq)$

Now balance the reaction.

Start by counting each atom.

reactant atoms	product atoms
1 Pb atom	1 Pb atom
1 Cl atom	2 Cl atoms
6 O atoms	3 O atoms
1 Na atoms	1 Na atoms
2 N atom	1 N atom

Since there are 2 Cl atoms on the right and only one Cl atom on the left, start by multiplying the NaCl on the left by 2 (2NaCl). Recount the atoms.

reactant atoms	product atoms
1 Pb atom	1 Pb atom
2 Cl atom	2 Cl atoms
6 O atoms	3 O atoms
2 Na atoms	1 Na atoms
2 N atom	1 N atom

The Cl atoms are balanced but there are 2 Na on the left and 1 Na on the right. Multiply $NaNO_3$ by 2 ($2NaNO_3$) and recount.

reactant atoms	product atoms
1 Pb atom	1 Pb atom
2 Cl atom	2 Cl atoms
6 O atoms	**6** O atoms
2 Na atoms	**2** Na atoms
2 N atom	**2** N atom

The atoms are now the same on both sides and you can write the balanced equation.

$$Pb(NO_3)_2(aq) + 2NaCl(aq) \rightarrow PbCl_2(s) + 2NaNO_3(aq)$$

6.77   *a. Complete and balance the precipitation reaction.*

$Na_2CO_3 + Pb(NO_3)_2 \rightarrow$

Use the procedure in previous problems to predict the products and to balance the equation. $PbCO_3$ is an insoluble product whereas $NaNO_3$ is a water-soluble product:

$$Na_2CO_3(aq) + Pb(NO_3)_2(aq) \rightarrow 2NaNO_3(aq) + PbCO_3(s)$$

b. Using Table 6.3, come up with your own precipitation reaction.

Show the formation of any insoluble salt, *e.g.*,

$$2KOH(aq) + MgCl_2(aq) \rightarrow Mg(OH)_2(s) + 2KCl(aq)$$

6.79 *Write the ionic equation and the net ionic equation for the reaction in Problem 6.73.*

The reaction in Problem 6.73 is:

$$CuSO_4(aq) + Na_2S(aq) \rightarrow CuS(s) + Na_2SO_4(aq)$$

In an ionic equation electrolytes are represented as individual ions. From the balanced equation above, any ionic compound that is in aqueous form is an electrolyte which can be written as individual ions. Because the product CuS is a solid that does not dissolve in water, this product is written in the original form.

Ionic equation:

$$Cu^{2+}(aq) + SO_4^{2-}(aq) + 2Na^+(aq) + S^{2-}(aq) \rightarrow CuS(s) + SO_4^{2-}(aq) + 2Na^+(aq)$$

The net ionic equation is derived from the ionic equation by removing any ions that appear unchanged between the reactant side and the product side. In this case, both the SO_4^{2-} and the Na^+ are unchanged and are considered spectator ions that can be removed from the equation. The reaction that actually takes place is the formation of solid barium sulfate from its constituent ions:

Net ionic equation:

$$\mathbf{Cu^{2+}(aq) + S^{2-}(aq) \rightarrow CuS(s)}$$

6.81 *Write the ionic equation and the net ionic equation for each reaction in Problem 6.75.*

a.   Balanced equation:
$$CaCl_2(aq) + Li_2CO_3(aq) \rightarrow CaCO_3(s) + 2LiCl(aq)$$

Ionic equation:
$$Ca^{2+}(aq) + 2Cl^-(aq) + 2Li^+(aq) + CO_3^{2-}(aq) \rightarrow CaCO_3(s) + 2Li^+(aq) + 2Cl^-(aq)$$

Net ionic equation:
$Ca^{2+}(aq) + CO_3^{2-}(aq) \rightarrow CaCO_3(s)$

b.   Balanced equation:
$Pb(NO_3)_2(aq) + 2NaCl(aq) \quad \rightarrow \quad PbCl_2(s) \ + \ 2NaNO_3(aq)$

Ionic equation:
$Pb^{2+}(aq) + 2NO_3^-(aq) + 2Na^+(aq) + 2Cl^-(aq) \rightarrow PbCl_2(s) + 2Na^+(aq) + 2NO_3^-(aq)$

Net ionic equation:
$Pb^{2+}(aq) + 2Cl^-(aq) \rightarrow PbCl_2(s)$

c.   Balanced equation:
$3CaBr_2(aq) + 2K_3PO_4(aq) \quad \rightarrow \quad Ca_3(PO_4)_2(s) \ + \ 6KBr(aq)$

Ionic equation:
$3Ca^{2+}(aq) + 6Br^-(aq) + 6K^+(aq) + 2PO_4^{3-}(aq) \rightarrow Ca_3(PO_4)_2(s) + 6K^+(aq) + 6Br^-(aq)$

Net ionic equation:
$3Ca^{2+}(aq) + 2PO_4^{3-}(aq) \rightarrow Ca_3(PO_4)_2(s)$

6.83   *What happens to the solubility of carbon dioxide gas (CO_2) in water in each situation? (Answer as increase, decrease, or no change.)*

a. The pressure of CO_2 over the solution is increased.

Increases. The solubility of a gas in a liquid increases as the pressure of the gas above the liquid increases.

b. The temperature is increased.

Decreases. The solubility of a gas in a liquid decreases as the temperature of the liquid increases.

6.85   *Two unopened bottles of carbonated water are at the same temperature. If one is opened at the top of a mountain and the other at sea level, which will produce more bubbles? Explain.*

The bottle opened at the top of the mountain will produce more bubbles. The solubility of a gas in a liquid decreases and, since the atmospheric pressure is lower at the top of the mountain, more gas will bubble out of solution.

6.87 *When your body metabolizes amino acids, one of the final end products is urea, a water-soluble compound that is removed from the body in urine. Why is urea soluble in water, when hexanamide, a related compound, is not?*

$$H_2N-\overset{\overset{\displaystyle O}{\|}}{C}-NH_2 \qquad\qquad CH_3-CH_2-CH_2-CH_2-CH_2-\overset{\overset{\displaystyle O}{\|}}{C}-NH_2$$

Urea   Hexanamide

Urea has more atoms capable of forming hydrogen bonds with water and it does not contain the nonpolar chain of carbon atoms present in hexanamide.

6.89 *Explain why CH₃CH₂OCH₂CH₃ is less soluble in water than its constitutional isomer CH₃CH₂CH₂CH₂OH.*

Molecules of $CH_3CH_2CH_2CH_2OH$ are able to form more hydrogen bonds with water.

6.91   *Vitamin D is produced in the skin upon exposure to sunlight. Based on its structure, is vitamin D hydrophilic, hydrophobic, or amphipathic?*

Vitamin D

Hydrophobic

First recall what each term means; hydrophilic—soluble in water, hydrophobic—insoluble in water, and amphipathic— has both hydrophilic and hydrophobic parts. Note that the vitamin D has an OH group on one end that would make it somewhat like water. However, the large nonpolar part of the molecule is dominant. That makes the compound hydrophobic.

6.93   *Explain how you would prepare a saturated aqueous solution of baking soda (NaHCO₃).*

A saturated solution is a solution that contains the maximum amount of solute that can be dissolved at a given temperature.

To prepare a saturated aqueous solution of $NaHCO_3$, add solid $NaHCO_3$ to water and mix to dissolve all of the $NaHCO_3$. Keep adding $NaHCO_3$ until no more of the solid dissolves.

6.95   *Albumin, a protein, is present in normal blood serum at concentrations of 3.5-5.5 g/dL. What is the % (w/v) of albumin in serum that contains 4.0 g of albumin per deciliter?*

Weight/volume percent is defined as:

$$\text{weight / volume percent} = \frac{\text{g solute}}{\text{mL of solution}} \times 100$$

To calculate the % (w/v) of a solution containing 4.0 g of albumin per deciliter of solution, first convert deciliter to milliliter:

$$1 \; \cancel{dL} \; \times \; \frac{1 \; \cancel{L}}{10 \; \cancel{dL}} \; \times \; \frac{1000 \; mL}{1 \; \cancel{L}} \; = \; 100 \; mL$$

$$weight / volume \; percent \; = \; \frac{4.0 \; g \; albumin}{100 \; mL \; of \; solution} \; \times \; 100 \; = \; 4.0 \; \% \; (w/v)$$

6.97   *If 100 mL of blood serum contains 5.0 mg of thyroxine, a hormone released by the thyroid gland, thyroxine levels are within the normal range for an adult. Express this concentration of thyroxine in parts per million and parts per billion.*

Parts per million and parts per billion are concentration units used to describe dilute solutions. The definitions of these two concentration quantities are as follows:

$$parts \; per \; million \; = \; \frac{g \; of \; solute}{mL \; of \; solution} \; \times \; 10^6$$

$$parts \; per \; billion \; = \; \frac{g \; of \; solute}{mL \; of \; solution} \; \times \; 10^9$$

Use the above equations to calculate the concentrations in parts per million and in parts per billion. First, convert milligrams of thyroxine to grams:

$$5.0 \; mg \; \times \; \frac{1 \; g}{1000 \; mg} \; = \; 5.0 \times 10^{-3} \; g \; thyroxine$$

$$parts \; per \; million \; = \; \frac{5.0 \times 10^{-3} \; g \; thyroxine}{100 \; mL \; of \; solution} \; \times \; 10^6 \; = \; 5.0 \times 10^1 \; ppm$$

$$parts \; per \; billion \; = \; \frac{5.0 \times 10^{-3} \; g \; thyroxine}{100 \; mL \; of \; solution} \; \times \; 10^9 \; = \; 5.0 \times 10^4 \; ppb$$

149

6.99   *Calculate the molarity of each.*

Molarity is equal to the moles of solute divided by the liters of solution.

a.   *0.33 mol of NaCl in 2.0 L of solution*

$$\text{molarity} \quad = \quad \frac{0.33 \text{ mol NaCl}}{2.0 \text{ L}} \quad = \quad 0.17 \text{ M}$$

b. *55.0 g of NaCl in 125 mL of solution*

First, convert grams of NaCl to moles of NaCl:

$$55.0 \text{ g NaCl} \quad \times \quad \frac{1 \text{ mol}}{58.5 \text{ g NaCl}} \quad = \quad 0.94 \text{ mol NaCl}$$

Convert the volume from mL to L:

$$125 \text{ mL} \quad \times \quad \frac{1 \text{ L}}{1000 \text{ mL}} \quad = \quad 0.125 \text{ L}$$

Then, proceed to calculate the molarity:

$$\text{molarity} \quad = \quad \frac{0.94 \text{ mol NaCl}}{0.125 \text{ L}} \quad = \quad 7.5 \text{ M}$$

6.101   *If 0.30 mg of KCl is present in 1.0 L of aqueous solution, what is the concentration in terms of the following?*

a.   *molarity*

Molarity is a concentration unit defined as the moles of solute per liter of solution. To calculate the molarity of the given solution, first, we need to calculate the moles of KCl in the solution:

$$0.30 \text{ mg KCl} \quad \times \quad \frac{1 \text{ g}}{1000 \text{ mg}} \quad = \quad 3.0 \times 10^{-4} \text{ g KCl}$$

$$3.0 \times 10^{-4} \text{ g KCl} \quad \times \quad \frac{1 \text{ mol KCl}}{74.6 \text{ g KCl}} \quad = \quad 4.0 \times 10^{-6} \text{ mol KCl}$$

$$\text{molarity} = \frac{4.0 \times 10^{-6} \text{ mol KCl}}{1.0 \text{ L}} = 4.0 \times 10^{-6} \text{ M}$$

b. *weight/volume percent*

$1.0 \text{ L} = 1.0 \times 10^3 \text{ mL}$

$$\text{weight / volume percent} = \frac{\text{g solute}}{\text{mL of solution}} \times 100$$

$$\text{weight / volume percent} = \frac{3.0 \times 10^{-4} \text{ g KCl}}{1.0 \times 10^3 \text{ mL of solution}} \times 100 = 3.0 \times 10^{-5} \% \text{ (w / v)}$$

c. *parts per thousand*

$$\text{parts per thousand} = \frac{3.0 \times 10^{-4} \text{ g KCl}}{1.0 \times 10^3 \text{ mL of solution}} \times 10^3 = 3.0 \times 10^{-4} \text{ ppt}$$

d. *parts per million*

$$\text{parts per million} = \frac{3.0 \times 10^{-4} \text{ g KCl}}{1.0 \times 10^3 \text{ mL of solution}} \times 10^6 = 3.0 \times 10^{-1} \text{ ppm}$$

e. *parts per billion*

$$\text{parts per billion} = \frac{3.0 \times 10^{-4} \text{ g KCl}}{1.0 \times 10^3 \text{ mL of solution}} \times 10^9 = 3.0 \times 10^2 \text{ ppb}$$

6.103 *For women, normal levels of uric acid in blood serum range from 26 to 60 ppm. If a female patient has 1.2 mg of uric acid in 10.0 mL of blood serum, is she within the normal range?*

First, calculate the ppm concentration of uric acid in the female's blood serum:

$$1.2 \; \cancel{mg} \text{ uric acid} \; \times \; \frac{1 \text{ g}}{1000 \; \cancel{mg}} \; = \; 1.2 \times 10^{-3} \text{ g uric acid}$$

$$\text{parts per million} \; = \; \frac{1.2 \times 10^{-3} \text{ g uric acid}}{10.0 \text{ mL of solution}} \; \times \; 10^{6} \; = \; 1.2 \times 10^{2} \text{ ppm}$$

The concentration of uric acid in the female's blood serum is not within the normal range (26 to 60 ppm). It is higher than normal.

6.105   *Thyroxine, a thyroid hormone, is present in normal blood serum at 58-167 nmol/L. What is the molar concentration of thyroxine in serum that contains 150 nmol of thyroxine per liter?*

Since molarity is defined as mol per liter, convert nmol to mol.

$$\frac{150 \; \cancel{nmol}}{L} \; \text{x} \; \frac{1 \times 10^{-9} \text{ mol}}{\cancel{nmol}} \; = \; 1.5 \times 10^{-7} \text{ mol/L}$$

1.5×10^{-7} mol thyroxine /L or 1.5×10^{-7} M thyroxine

6.107   *The serum concentration of cortisol, a hormone, is expected to be in the range 8-20 mg/dL. What is the part per million concentration of cortisol in serum that contains 10 mg of cortisol per deciliter?*

Parts per million is the weight of solute (in grams) divided by the volume (in mL) of the solution multiplied by 10^{6}. Convert mg to g and dL to mL, then multiply by 10^{6}.

$$\frac{10 \; \cancel{mg}}{\cancel{dL}} \; \text{x} \; \frac{1 \times 10^{-3} \text{ g}}{\cancel{mg}} \; \text{x} \; \frac{1 \; \cancel{dL}}{1 \times 10^{2} \text{ mL}} \; \text{x} \; 10^{6} \; = \; 100 \text{ ppm}$$

100 ppm cortisol

6.109 *The normal serum concentration of potassium ion (K^+) is 3.5-4.9 mEq/L. Convert this concentration range into mmol/L.*

For an ion with a 1+ charge, 1 mol = 1 Eq, so 1 mmol = 1 mEq.

$$\frac{3.5-4.9 \; \text{mEq}}{L} \; \text{x} \; \frac{1 \; \text{mmol}}{\text{mEq}} = 3.5\text{-}4.9 \; \text{mmol/L}$$

6.111 *The normal serum concentration of chloride ion (Cl^-) is 95-107 mmol/L. Convert this concentration range into mEq/L.*

For an ion with a 1- charge 1 mol = 1 Eq, so 1 mmol = 1 mEq.

$$\frac{95-107 \; \text{mmol}}{L} \; \text{x} \; \frac{1 \; \text{mEq}}{\text{mmol}} = 95-107 \; \text{mEq/L}$$

6.113 *How many milliequivalents of bicarbonate (HCO_3^-) are present in a 75.0 mL blood serum sample with a concentration of 25 mEq/L of HCO_3^-?*

Convert mL to L and multiply by the concentration in mEq/L.

$$75.0 \; \text{mL} \; \text{x} \; \frac{1 \; \text{x} \; 10^{-3} \; L}{\text{mL}} \; \text{x} \; \frac{25 \; \text{mEq}}{1 \; L} = 1.9 \; \text{mEq}$$

1.9 mEq HCO_3^-

6.115 *How many moles of sodium ion (Na^+) are present in a 10.0 mL blood serum sample with a concentration of 132 mEq/L?*

Convert mL to L and mmol to mEq (for Na^+, 1 mmol = mEq) and then convert mmol to mol.

$$10.0 \; \text{mL} \; \text{x} \; \frac{1 \; \text{x} \; 10^{-3} \; L}{\text{mL}} \; \text{x} \; \frac{132 \; \text{mEq}}{1 \; L} \; \text{x} \; \frac{1 \; \text{mmol}}{1 \; \text{mEq}} \; \text{x} \; \frac{1 \; \text{x} \; 10^{-3} \; \text{mol}}{1 \; \text{mmol}} = 1.32 \; \text{x} \; 10^{-3} \; \text{mol}$$

1.32 x 10^{-3} mol Na^+

6.117   *If 15.0 mL of 3.0 M HCl are diluted to a final volume of 100.0 mL, what is the new concentration?*

The dilution equation is $V_{original} \times C_{original} = V_{final} \times C_{final}$. To use this equation it is not necessary to have a particular volume or concentration unit, provided that both volume units are the same and that both concentration units are the same. The answer will have the same unit as the value that is given in the problem.

$$V_{original} \times C_{original} = V_{final} \times C_{final}$$

$$C_{final} = \frac{V_{original} \times C_{original}}{V_{final}}$$

$$C_{final} = \frac{15.0 \text{ mL} \times 3.0 \text{ M}}{100.0 \text{ mL}} = 0.45 \text{ M HCl}$$

6.119   *A 10.0% (w/v) solution of ethyl alcohol is diluted from 50.0 mL to 200.0 mL. What is the new weight/volume percent?*

Recall the dilution equation is $V_{original} \times C_{original} = V_{final} \times C_{final}$.

$$C_{final} = \frac{V_{original} \times C_{original}}{V_{final}}$$

$$C_{final} = \frac{50.0 \text{ mL} \times 10.0\%}{200.0 \text{ mL}} = 2.50\% \text{ (w/v)}$$

6.121   *How many milliliters of 2.00 M NaOH are needed to prepare 300.0 mL of 1.50 M NaOH?*

Recall the dilution equation is $V_{original} \times C_{original} = V_{final} \times C_{final}$.

$$V_{original} = \frac{V_{final} \times C_{final}}{C_{original}}$$

$$V_{original} = \frac{300.0 \text{ mL} \times 1.50 \text{ M}}{2.00 \text{ M}} = 225 \text{ mL}$$

6.123   *Calculate the final volume required to prepare each solution.*

a. Starting with 100.0 mL of 1.00 M KBr, prepare 0.500 M KBr.

Recall the dilution equation is $V_{original} \times C_{original} = V_{final} \times C_{final}$.

$$V_{final} = \frac{V_{original} \times C_{original}}{C_{final}}$$

$$V_{final} = \frac{100.0 \text{ mL} \times 1.00 \cancel{M}}{0.500 \cancel{M}} = 200. \text{ mL}$$

b. Starting with 50.0 mL of 0.250 M alanine (an amino acid), prepare 0.110 M alanine.

Recall the dilution equation is $V_{original} \times C_{original} = V_{final} \times C_{final}$.

$$V_{final} = \frac{V_{original} \times C_{original}}{C_{final}}$$

$$V_{final} = \frac{50.0 \text{ mL} \times 0.250 \cancel{M}}{0.110 \cancel{M}} = 114 \text{ mL}$$

6.125   *How is a colloid different from a suspension?*

Suspensions and colloids are special types of mixtures. In a suspension, the mixture is composed of large particles suspended in a liquid. In a colloid, the mixture contains particles that are smaller than particles in a suspension. The smaller size of the particles in a colloid differentiates them from a suspension. The particles in a colloid are not heavy enough to settle out in the liquid upon standing.

6.127   *What do you end up with if you pour dirt into water and stir – a solution, a colloid, or a suspension?*

When dirt is mixed into water by stirring, a suspension is made because the particles of dirt are heavy enough to separate out upon standing.

6.129   *a. Define the term diffusion*

Diffusion is the movement of substances from areas of higher concentration to areas of lower concentration.

b. A process called active transport moves certain ions and compounds across cell membranes from areas of lower concentration to areas of higher concentration. Does active transport involve diffusion? Explain

No. In active transport solutes move in a direction opposite to diffusion.

6.131   *To make pickles, you soak cucumbers in a concentrated salt solution called brine. Describe how this process is related to osmosis.*

Osmosis is the net flow of water across a membrane from a solution of lower concentration to a solution of higher concentration. Brine, concentrated salt solution, has a higher concentration of salt than the liquid in cucumber. As a result, there will be a net flow of water from the cucumber into the brine, and a pickle results.

6.133   *During severe bleeding, ADH (a hormone released by the hypothalamus) causes vasoconstriction (shrinking of the blood vessels) to take place. What effect does a decrease in blood vessel volume have on blood pressure?*

A decrease in blood vessel volume causes an increase in blood pressure.

6.135   *The chest compressions given during cardiopulmonary resuscitation (CPR) cause the injured person to exhale. Explain why, in terms of Boyle's law.*

Chest compressions cause the volume of the lungs to decrease. According to Boyle's law, if the volume of the lungs and the gas inside decreases, the pressure of the gas will increase. The gas will then move from the lungs (higher pressure) to the atmosphere that surrounds you (lower pressure).

6.137 *Why is the prodrug chloramphenicol palmitate (Figure 6.24) less water soluble than chloramphenicol?*

The long hydrocarbon chain in chloramphenicol palmitate makes the molecule less polar and therefore less soluble in water.

6.139 *Saliva has some characteristics of a solution. Explain.*

Saliva is a homogenous mixture of ionic compounds, proteins, and nucleic acids.

6.141 *Why is it important that dialyzing solution be isotonic with blood?*

Dialyzing solution has to be isotonic with blood to ensure that there is no net flow of essential solutes into or out of the blood.

6.143 *a. A 1.50 L gas cylinder contains a mixture of nitrogen gas (N_2) and hydrogen gas (H_2) at a temperature of $35^{\circ}C$ and a pressure 895 torr. What is the pressure in atmosphere (atm)?*

1 atm = 760 torr or mmHg = 14.7 psi

$$895 \; \text{torr} \; \times \; \frac{1 \; \text{atm}}{760 \; \text{torr}} \; = \; 1.18 \; \text{atm}$$

b. What is the pressure in pounds per square inch (psi)?

$$895 \; \text{torr} \; \times \; \frac{14.7 \; \text{psi}}{760 \; \text{torr}} \; = \; 17.3 \; \text{psi}$$

c. If the partial pressure of N_2 is 615 torr, what is the partial pressure of H_2?

P_{H2} = 895 torr - 615 torr = 280. torr or 2.80×10^2 torr

d. If the temperature is increased to $70.0^{\circ}C$, what is the new partial pressure of each gas and what is the new total pressure?

New partial pressure of N_2:
P_1 = 615 torr P_2 = ?
T_1 = 35°C + 273.15 = 308 K T_2 = 70.0°C + 273.15 = 343.2 K

157

New P_{N2} $= \dfrac{615 \text{ torr}}{308 \text{ K}}$ x 343.2 K = 685 torr

New partial pressure of H_2:

$P_1 = 2.80 \times 10^2$ torr $P_2 = ?$

$T_1 = 35°C + 273.15 = 308$ K $T_2 = 70.0°C + 273.15 = 343.2$ K

New P_{H2} $= \dfrac{2.80 \times 10^2 \text{ torr}}{308 \text{ K}}$ x 343.2 K = 312 torr

New total pressure = 685 torr + 312 torr = 997 torr

e. *How many moles of N_2 and moles of H_2 are present in the cylinder?*

P_{N2} = 615 ~~torr~~ x $\dfrac{1 \text{ atm}}{760 \text{ ~~torr~~}}$ = 0.809 atm

P_{H2} = 280. ~~torr~~ x $\dfrac{1 \text{ atm}}{760 \text{ ~~torr~~}}$ = 0.368 atm

n_{N2} $= \dfrac{0.809 \text{ ~~atm~~} \times 1.50 \text{ ~~L~~}}{0.0821 \text{ ~~L~~}\cdot\text{~~atm~~}/\text{mol}\cdot\text{~~K~~} \times 308 \text{ ~~K~~}}$ = 0.0480 mol N_2

n_{H2} $= \dfrac{0.368 \text{ ~~atm~~} \times 1.50 \text{ ~~L~~}}{0.0821 \text{ ~~L~~}\cdot\text{~~atm~~}/\text{mol}\cdot\text{~~K~~} \times 308 \text{ ~~K~~}}$ = 0.0219 mol H_2

f. *How many grams of N_2 and grams of H_2 are present in the cylinder?*

0.0480 ~~mol N_2~~ x $\dfrac{28.01 \text{ g } N_2}{1 \text{ ~~mol N_2~~}}$ = 1.34 g N_2

0.0219 ~~mol H_2~~ x $\dfrac{2.02 \text{ g } H_2}{1 \text{ ~~mol H_2~~}}$ = 0.0442 g H_2

g. *How many molecules of N_2 and molecules of H_2 are present in the cylinder?*

$$0.0480 \; \text{mol N}_2 \times \frac{6.02 \times 10^{23} \; \text{molecules N}_2}{1 \; \text{mol N}_2} = 2.89 \times 10^{22} \; \text{molecules N}_2$$

$$0.0219 \; \text{mol H}_2 \times \frac{6.02 \times 10^{23} \; \text{molecules H}_2}{1 \; \text{mol H}_2} = 1.32 \times 10^{22} \; \text{molecules H}_2$$

h. *Nitrogen gas and hydrogen gas react according to the equation:*

$$N_2(g) + 3H_2(g) \rightarrow 2NH_3(g)$$

Assuming that conditions are right for the reaction to take place, which is the limiting reactant, N_2 or H_2?

Calculate the mol of H_2 required to react with all of the N_2 present:

$$0.0480 \; \text{mol N}_2 \times \frac{3 \; \text{mol H}_2}{1 \; \text{mol N}_2} = 0.144 \; \text{mol H}_2$$

Because this is more than the amount of H_2 available, not all of the N_2 will react. H_2 is the limiting reactant.

i. *For the reaction in part h, what is the theoretical yield (in grams) of NH_3*

$$0.0219 \; \text{mol H}_2 \times \frac{2 \; \text{mol NH}_3}{3 \; \text{mol H}_2} \times \frac{17.03 \; \text{g NH}_3}{1 \; \text{mol NH}_3} = 0.249 \; \text{g NH}_3$$

6.145 *When given to a patient, the drug allopurinol is converted to oxypurinol. Oxypurinol blocks the action of xanthine oxidase, an enzyme that catalyzes one of the steps in the production of uric acid from purines.*

Allopurinol   Oxypurinol

a. *Is allopurinol oxidized or reduced when it is converted to oxypurinol?*

Oxidized. Recall from Chapter 6 that an atom is reduced if it loses oxygen and/or gains hydrogen. It is oxidized if it gains oxygen and/or loses hydrogen. In this reaction, the allopurinol carbon that undergoes change is losing a hydrogen atom and gaining an OH. This means that it is oxidized.

b. Might allopurinol be considered a prodrug? (Refer to Health Link: Prodrugs.)

Yes. Allopurinol itself is inactive, but it is converted to an active compound in the body.

c. Suggest an explanation for the fact that oxypurinol is better at blocking xanthine oxidase than is allopurinol. (Hint: look at the structures in the reactions presented just before the chapter summary.)

The structure of oxypurinol more closely resembles that of xanthine, so it is more likely to interact with xanthine oxidase and inhibit its enzymatic ability.

6.147 *How many grams of solute are needed to make 250 mL of 0.100 M K_2SO_4?*

First, convert 250 mL to L then use the molarity concentration as a conversion factor to calculate the number of moles of K_2SO_4. Convert to grams of K_2SO_4 to get the final answer.

$$250 \; \cancel{mL} \; \times \frac{1 \; L}{1000 \; \cancel{mL}} = \; 0.250 \; L$$

$$0.250 \; \cancel{L} \; \times \; \frac{0.100 \; mole \; K_2SO_4}{1 \; \cancel{L}} = \; 0.0250 \; mole \; K_2SO_4$$

$$0.0250 \; \cancel{mole \; K_2SO_4} \; \times \; \frac{174.3 \; g \; K_2SO_4}{1 \; \cancel{mole \; K_2SO_4}} = \; 4.36 \; g \; K_2SO_4$$

Chapter 7
Acids, Bases, and Equilibrium

Solutions to Problems

7.1   *a. Write the equilibrium constant expression for the reaction.*

$$A \rightleftharpoons B$$

$$K = \frac{[B]}{[A]}$$

b. For this reaction $K_{eq} = 2$. The drawing below represents an equilibrium mixture of A and B. Explain

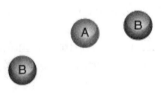

As shown in the drawing above, the ratio of the concentration of B to A, [B]/[A], is equal to 2. Because this is the same values as the equilibrium constant,

$$K = \frac{[B]}{[A]} = 2 \text{, then this reaction mixture must be at equilibrium.}$$

c. If 3 additional Bs are added, the mixture is no longer in equilibrium. Explain.

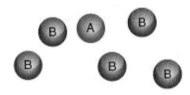

When the concentration of B is increased, the rate of the reverse reaction becomes greater than the forward rate. The mixture is no longer in equilibrium because the reverse rate and the forward rate are no longer equal.

d. To adjust to equilibrium, which will take place for the mixture in part c: a net forward reaction or a net reverse reaction?

A net reverse reaction will take place to consume the additional B's until the ratio of [B]/[A] goes back down to 2. When K = 2, the reaction is at equilibrium.

e. Based on your answer to part d, redraw the picture in part c as it will appear once the mixture reaches equilibrium.

One B is converted to one A, decreasing the number of B's to 4 and increasing the number of A's to 2, bringing the ratio back to its equilibrium value:

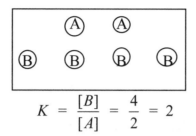

$$K = \frac{[B]}{[A]} = \frac{4}{2} = 2$$

7.3   *Which of the following observations would indicate that a compound might be an acid?*

a. Turns litmus pink
b. Turns litmus blue
c. Has a bitter taste
d. Has a sour taste
e. Dissolves a metal
f. Feels slippery

The observations a, d, and e are all properties of an acid. The others (b, c, and f) are properties of a base.

7.5   *Name each of the following as an acid and as a binary compound.*

Covalent binary compounds are named by using the name of the first element and then naming the second element with its end changed to "- *ide*". When binary compounds are named as acids, the ending is changed to *"-ic acid"* or *"-ous acid."*

a.   *HCl*   hydrochloric acid   hydrogen chloride

b.   *HBr*   hydrobromic acid   hydrogen bromide

7.7   *Give the formula for the conjugate base of each acid.*

The conjugate base of an acid is the compound or ion that is formed after the acid loses an H^+. When an H^+ is removed, the resulting particle gains a negative charge.

	acid	conjugate base
a.	H_2CO_3	HCO_3^-
b.	HCO_3^-	CO_3^{2-}
c.	H_2SO_4	HSO_4^-
d.	$H_2PO_4^-$	HPO_4^{2-}

7.9 *Name the conjugate base of each acid in Problem 7.7*

Each of these conjugate bases is a polyatomic ion.

conjugate base

a.	HCO_3^-	hydrogen carbonate ion
b.	CO_3^{2-}	carbonate ion
c.	HSO_4^-	hydrogen sulfate ion
d.	HPO_4^{2-}	hydrogen phosphate ion

7.11 *Give the formula for the conjugate acid of each base.*

The conjugate acid of a base is the compound or ion that is formed after the base gains an H^+. When an H^+ is added, the resulting particle gains a positive charge.

	base	conjugate acid
a.	F^-	HF
b.	CN^-	HCN
c.	NH_3	NH_4^+
d.	NO_2^-	HNO_2

7.13   *Water is amphoteric (can act as an acid or a base).*

   a. *Write a chemical equation that shows water acting as an acid.*

   H_2O donates an H^+ to NH_3:

   $$H_2O + NH_3 \rightleftharpoons OH^- + NH_4^+$$

   b. *In this reaction, what is the conjugate base of water?*

   OH^-. The conjugate base of H_2O, OH^-, is what is left after water donates an H^+ to NH_3.

7.15   *HCO_3^- is amphoteric.*

   a. *Write the equation for the reaction that takes place between HCO_3^- and the acid HCl.*

   The acid HCl donates H^+ to the base HCO_3^-.

   $$HCO_3^-\,(aq) + HCl(aq) \rightarrow H_2CO_3(aq) + Cl^-(aq)$$

   b. *Write the equation for the reaction that takes place between HCO_3^- and the base OH^-.*

   The acid HCO_3^- donates H^+ to the base OH^-.

   $$HCO_3^-\,(aq) + OH^-\,(aq) \rightarrow CO_3^{2-}\,(aq) + H_2O(l)$$

7.17   *Identify the Bronsted-Lowry acids and bases for the forward and reverse reactions of each.*

   a. $F^-(aq) + HCl\,(aq) \rightleftharpoons HF(aq) + Cl^-\,(aq)$

   Bronsted-Lowry acids donate H^+ and Bronsted-Lowry bases accept H^+. In the forward reaction, HCl gives up H^+ to F^-. In the reverse reaction, HF donates H^+ to Cl^-.

   $$F^-(aq) \;+\; HCl\,(aq) \rightleftharpoons HF(aq) \;+\; Cl^-\,(aq)$$
   base   acid   acid   base

   b. $CH_3CO_2H\,(aq) + NO_3^-\,(aq) \rightleftharpoons CH_3CO_2^-\,(aq) + HNO_3\,(aq)$

Bronsted-Lowry acids donate H^+ and Bronsted-Lowry bases accept H^+. In the forward reaction, CH_3CO_2H donates H^+ to NO_3^-. In the reverse reaction, HNO_3 donates H^+ to $CH_3CO_2^-$.

$$CH_3CO_2H \ (aq) \ + \ NO_3^- \ (aq) \rightleftharpoons CH_3CO_2^- \ (aq) \ + \ HNO_3 \ (aq)$$

| acid | base | base | acid |

7.19   *Complete each acid-base reaction. For the forward and reverse reactions, identify each acid and its conjugate base.*

An acid-base reaction involves the transfer of a H^+ ion from the acid to the base. The acid is the ion or molecule that donates the H^+ ion and the base is the ion or molecule that accepts the H^+ ion. An acid and its conjugate base differ only by one H^+ ion.

a. NH_4^+ $+$ SO_4^{-2} $\rightleftharpoons$

$NH_4^+ \ + \ SO_4^{-2} \rightleftharpoons NH_3 \ + \ HSO_4^-$
acid   conjugate   conjugate   acid
                base   base

b. CN^- $+$ HI

$CN^- \ + \ HI \rightleftharpoons HCN \ + \ I^-$
conjugate   acid   acid   conjugate
base   base

7.21   *Which of the following statements are correct at equilibrium?*
a. The concentration of reactants is always equal to the concentration of products.
b. No reactants are converted into products.
c. The rate of the forward reaction is equal to the rate of the reverse reaction.

Only *c.* is correct. At equilibrium, the rates of the forward and reverse reactions are equal. Depending on the reaction, the equilibrium concentration of reactants can be greater than or less than product concentrations, therefore *a.* is not true. Answer *b.* is not true because at equilibrium the reaction continues to take place in both directions.

7.23 *Balance the chemical equation and write the equilibrium constant expression.*

$$CH_4 \quad + \quad H_2O \quad \rightleftharpoons \quad H_2 \quad + \quad CO$$

For the generalized reaction, $aA + bB \rightleftharpoons cC + dD$

the equilibrium constant expression is written:

$$K_{eq} = \frac{[C]^c[D]^d}{[A]^a[B]^b}$$

The balanced equation is:

$$CH_4 \quad + \quad H_2O \quad \rightleftharpoons \quad 3H_2 \quad + \quad CO$$

and the equilibrium constant expression is:

$$K_{eq} = \frac{[H_2]^3[CO]}{[CH_4][H_2O]}$$

7.25 *Balance the chemical equation and write the equilibrium constant expression.*

$$SO_2 \quad + \quad O_2 \quad \rightleftharpoons \quad SO_3$$

The balanced equation is:

$$2SO_2 \quad + \quad O_2 \quad \rightleftharpoons \quad 2SO_3$$

and the equilibrium constant expression is:

$$K_{eq} = \frac{[SO_3]^2}{[SO_2]^2[O_2]}$$

7.27 *Write the equilibrium constant expression for each reaction. In part b, H_2O is the solvent.*

a. $C(s) + CO_2(g) \rightleftharpoons 2CO(g)$

Remember that the concentration of solids and solvents are not included in K_{eq}.

$$K_{eq} = \frac{[CO]^2}{[CO_2]}$$

b. $NH_3(aq) + H_2O(l) \rightleftharpoons NH_4^+(aq) + OH^-(aq)$

Remember that the concentration of solids and solvents are not included in K_{eq}.

$$K_{eq} = \frac{[NH_4][OH^-]}{[NH_3]}$$

7.29   *Explain why the concentrations of solids do not appear in equilibrium constant expressions.*

Only concentrations that change should be included in the equilibrium constant. The concentration of a solid is constant at a given temperature.

7.31   *Write the reaction equation from which each equilibrium constant expression is derived (assume that no solids or solvents are present).*

The generalized equilibrium constant expression

$$K_{eq} = \frac{[C]^c[D]^d}{[A]^a[B]^b}$$

is based on the reaction equation
$$aA + bB \rightleftharpoons cC + dD$$

a. $K_{eq} = \dfrac{[NOCl]^2}{[NO]^2[Cl_2]}$

$2 NO + Cl_2 \rightleftharpoons 2 NOCl$

b. $K_{eq} = \dfrac{[HBr]^2}{[H_2][Br_2]}$

$H_2 + Br_2 \rightleftharpoons 2 HBr$

7.33 K_{eq} for the reaction below has a value of 4.2×10^{-4}.

$$CH_3NH_2(aq) + H_2O(l) \rightleftharpoons CH_3NH_3^+(aq) + OH^-(aq)$$

a.   Write the equilibrium constant expression for the reaction.

$$K_{eq} = \frac{[CH_3NH_3^+][OH^-]}{[CH_3NH_2]}$$

b.   Which are there more of at equilibrium, products or reactants?

Reactants. In the expression for K_{eq}, the reactant concentration is in the denominator. Since the K_{eq} value is very small this tells you that the reactant concentration is much larger than the product concentrations.

7.35   For each reaction, which are there more of at equilibrium, reactants or products?

a. $HPO_4^{2-} + CN^- \rightleftharpoons PO_4^{3-} + HCN$ $K_{eq} = 8.6 \times 10^{-4}$

Reactants. Because K_{eq} is small, reactant concentrations are larger than product concentrations.

b. $H_3PO_4 + CN^- \rightleftharpoons H_2PO_4^- + HCN$ $K_{eq} = 1.5 \times 10^7$

Products. Because K_{eq} is big, product concentrations are larger than reactant concentrations.

7.37   The enzyme carbonic anhydrase catalyzes the rapid conversion of CO_2 and H_2O into H_2CO_3.

$$CO_2(g) + H_2O(l) \rightleftharpoons H_2CO_3(aq)$$

a. Write the equilibrium constant expression for this reaction.

$$K_{eq} = \frac{[H_2CO_3]}{[CO_2]}$$

b. What effect, if any, does doubling the amount of carbonic anhydrase have on an equilibrium mixture of H_2CO_3, CO_2, and H_2O? Explain.

No effect. The catalyst increases the rate of the forward and reverse reactions to the same extent, but the concentrations of substances present at equilibrium are not affected.

7.39   *When carbon monoxide reacts with hydrogen gas, methanol (CH₃OH) is formed.*

$$CO(g) + 2H_2(g) \rightleftharpoons CH_3OH(g)$$

a. For the equilibrium above, what is the effect of increasing [CO]? Of decreasing [H₂]? Of increasing [CH₃OH]?

Increasing [CO] favors the forward reaction (makes more products). Decreasing [H₂] slows the rate of the forward reaction (favors the reverse reaction and makes more reactants). Increasing [CH₃OH] increases the rate of the reverse reaction (favors the reverse reaction and makes more reactants).

b. What would be the effect of continually removing CH₃OH from the reaction?

The reaction would continually move to make CH₃OH and would never reach equilibrium.

7.41   *One step in glycolysis (Section 14.4) involves the reversible conversion of glucose 6-phosphate into fructose 6-phosphate.*

Glucose 6-phosphate ⇌ fructose 6-phosphate

a. *For the equilibrium above, what is the effect of increasing the concentration of glucose 6-phosphate?*

There is a net forward reaction until equilibrium is reestablished.

b. *What is the effect of increasing the concentration of fructose 6-phosphate?*

There is a net reverse reaction until equilibrium is reestablished.

c. *What is the effect of decreasing the concentration of glucose 6-phospate?*

There is a net reverse reaction until equilibrium is reestablished.

7.43   *Calculate the H_3O^+ concentration present in water when*

To calculate the $[H_3O^+]$,

$$K_w = [H_3O^+][OH^-] \quad \text{and} \quad [H_3O^+] = \frac{K_w}{[OH^-]}$$

a. *[OH] = 8.4 x 10^{-3} M*

$$[H_3O^+] = \frac{1.0 \times 10^{-14}}{8.4 \times 10^{-3}} = 1.2 \times 10^{-12} \text{ M}$$

b. *[OH] = 2.9 x 10^{-9} M*

$$[H_3O^+] = \frac{1.0 \times 10^{-14}}{2.9 \times 10^{-9}} = 3.5 \times 10^{-6} \text{ M}$$

c. *[OH] = 5.8 x 10^{-1} M*

$$[H_3O^+] = \frac{1.0 \times 10^{-14}}{5.8 \times 10^{-1}} = 1.7 \times 10^{-14} \text{ M}$$

7.45   *In Problem 7.43, indicate whether each solution is acidic, basic, or neutral.*

A solution with $[H_3O^+] > 1.0 \times 10^{-7}$ M is acidic, with $[H_3O^+] < 1.0 \times 10^{-7}$ M is basic, and $[H_3O^+] = 1.0 \times 10^{-7}$ M is neutral.

a. *$[H_3O^+]$ = 1.2 x 10^{-12} M*   basic

b. *$[H_3O^+]$ = 3.5 x 10^{-6} M*   acidic

c. *$[H_3O^+]$ = 1.7 x 10^{-14} M*   basic

7.47   *Calculate the OH concentration present in water when*

$$[OH^-] = \frac{K_w}{[H_3O^+]}$$

a. *$[H_3O^+]$ = 9.1 x 10^{-7} M*

$$[OH^-] = \frac{1.0 \times 10^{-14}}{9.1 \times 10^{-7}} = 1.1 \times 10^{-8} \text{ M}$$

b. $[H_3O^+] = 1.3 \times 10^{-3}$ M

$$[OH^-] = \frac{1.0 \times 10^{-14}}{1.3 \times 10^{-3}} = 7.7 \times 10^{-12} \text{ M}$$

c. $[H_3O^+] = 8.8 \times 10^{-2}$ M.

$$[OH^-] = \frac{1.0 \times 10^{-14}}{8.8 \times 10^{-2}} = 1.1 \times 10^{-13} \text{ M}$$

7.49   *In Problem 7.47, indicate whether each solution is acidic, basic, or neutral.*

A solution with $[H_3O^+] > 1.0 \times 10^{-7}$ M is acidic, with $[H_3O^+] < 1.0 \times 10^{-7}$ M is basic, and $[H_3O^+] = 1.0 \times 10^{-7}$ M is neutral.

a. $[H_3O^+] = 9.1 \times 10^{-7}$ M   acidic

b. $[H_3O^+] = 1.3 \times 10^{-3}$ M   acidic

c. $[H_3O^+] = 8.8 \times 10^{-2}$ M   acidic

7.51   *Calculate the pH of a solution in which*

The pH of a solution can be calculated from the H_3O^+ concentration using the expression:
$$pH = -\log [H_3O^+]$$

On a standard scientific calculator, enter the value for $[H_3O^+]$ into the calculator, then press the **log** button. Multiply the result by -1. The number of digits after the decimal point in the pH value should be the same as the number of significant digits in the concentration value.

a. $[H_3O^+] = 1 \times 10^{-5}$ M

pH $= -\log (1 \times 10^{-5})$ $= 5.0$

b. $[H_3O^+] = 3.9 \times 10^{-2}$ M

pH $= -\log (3.9 \times 10^{-2})$ $= 1.41$

c. $[H_3O^+] = 1 \times 10^{-7}$ M

pH $= -\log (1 \times 10^{-7})$ $= 7.0$

d. $[H_3O^+] = 7.0 \times 10^{-5}$ M

pH $= -\log (7.0 \times 10^{-5})$ $= 4.15$

7.53 *In Problem 7.51, indicate whether each solution is acidic, basic, or neutral.*

In a neutral solution pH $= 7$, in an acidic solution pH < 7, and in a basic solution pH > 7.

a. pH $= 5.0$ acidic

b. pH $= 1.41$ acidic

c. pH $= 7.0$ neutral

d. pH $= 4.15$ acidic

7.55 *Calculate the pH of a solution in which*

a. $[OH] = 6.8 \times 10^{-7}$ M
b. $[OH] = 1 \times 10^{-7}$ M
c. $[OH^-] = 7.0 \times 10^{-5}$ M
d. $[OH^-] = 1 \times 10^{-1}$ M

First calculate the $[H_3O^+]$: $\quad [H_3O^+] = \dfrac{K_w}{[OH^-]} = \dfrac{1.0 \times 10^{-14}}{[OH^-]}$

Then calculate the pH: $\qquad\qquad$ pH $= -\log [H_3O^+]$

a. $[OH] = 6.8 \times 10^{-7}$ M

$$[H_3O^+] = \frac{1.0 \times 10^{-14}}{6.8 \times 10^{-7}} = 1.5 \times 10^{-8} \text{ M}$$

pH $= -\log (1.5 \times 10^{-8})$ $= 7.82$

172

b. $[OH^-] = 1 \times 10^{-7} M$

$$[H_3O^+] \quad = \quad \frac{1.0 \times 10^{-14}}{1 \times 10^{-7}} \quad = \quad 1 \times 10^{-7} \, M$$

$$pH \quad = -\log (1 \times 10^{-7}) \qquad = 7.0$$

c. $[OH^-] = 7.0 \times 10^{-5} M$

$$[H_3O^+] \quad = \quad \frac{1.0 \times 10^{-14}}{7.0 \times 10^{-5}} \quad = \quad 1.4 \times 10^{-10} \, M$$

$$pH \quad = -\log (1.4 \times 10^{-10}) \qquad = 9.85$$

d. $[OH^-] = 1 \times 10^{-1} M$

$$[H_3O^+] \quad = \quad \frac{1.0 \times 10^{-14}}{1 \times 10^{-1}} \quad = \quad 1 \times 10^{-13} \, M$$

$$pH \quad = -\log (1 \times 10^{-13}) \qquad = 13.0$$

7.57   *In problem 7.55, indicate whether each solution is acidic, basic, or neutral.*

In a neutral solution pH = 7, in an acidic solution pH < 7, and in a basic solution pH > 7.

    a.   pH = 7.82   basic

    b.   pH = 7.0   neutral

    c.   pH = 9.85   basic

    d.   pH = 13.0   basic

7.59   *What is the concentration of H_3O^+ in a solution if the pH is*
    a. 7.00   *c. 9.37*
    b. 1.74   *d. 10.3*

To calculate the concentration of H_3O^+, $[H_3O^+]$, from the pH, use the following equation:

$[H_3O^+] = 10^{-pH}$

a. 7.00

$$[H_3O^+] = 10^{-7.00} = 1.0 \text{ x } 10^{-7} \text{ M}$$

b. 1.74

$$[H_3O^+] = 10^{-1.74} = 1.8 \text{ x } 10^{-2} \text{ M}$$

c. 9.37

$$[H_3O^+] = 10^{-9.37} = 4.3 \text{ x } 10^{-10} \text{ M}$$

d. 10.3

$$[H_3O^+] = 10^{-10.3} = 5 \text{ x } 10^{-11} \text{ M}$$

7.61   *Alkali metals react with water to produce alkaline (basic) solutions. Na for example, reacts with water to form NaOH and H₂(g). Write a balanced chemical equation for this reaction.*

$$2Na(s) + 2H_2O(l) \rightarrow 2NaOH(aq) + H_2(g)$$

7.63   *In problem 7.61, what is reduced and what is oxidized?*

H is reduced because it lost a bond to oxygen. Na is oxidized because it lost electrons, going from neutral Na(s) to Na$^+$.

7.65   *A 1.00 mL sample of blood serum has a pH of 7.35.*

a.   What is the concentration of H₃O⁺?

$$[H_3O^+] = 10^{-pH}$$

$$[H_3O^+] = 10^{-7.45} = 4.5 \text{ x } 10^{-8} \text{ M}$$

b.   What is the concentration of OH⁻?

$$[OH^-] = 1.0 \text{ x } 10^{-14} / [H_3O^+]$$

$$[OH^-] = 1.0 \text{ x } 10^{-14} / 4.5 \text{ x } 10^{-8}$$

$$[OH^-] = 2.2 \text{ x } 10^{-7} \text{ M}$$

c. *How many moles of H_3O^+ are present?*

$$1.00 \, \cancel{mL} \; \times \; \frac{4.5 \times 10^{-8} \, mol}{1 \, \cancel{L}} \; \times \; \frac{1 \times 10^{-3} \, \cancel{L}}{\cancel{mL}} \; = \; 4.5 \times 10^{-11} \, mol$$

d. *How many moles of OH^- are present?*

$$1.00 \, \cancel{mL} \; \times \; \frac{2.2 \times 10^{-7} \, mol}{1 \, \cancel{L}} \; \times \; \frac{1 \times 10^{-3} \, \cancel{L}}{\cancel{mL}} \; = \; 2.2 \times 10^{-10} \, mol$$

e. *How many H_3O^+ ions are present?*

$$4.5 \times 10^{-11} \, \cancel{mol} \; \times \; \frac{6.02 \times 10^{23} \, ions}{1 \, \cancel{mol}} \; = \; 2.7 \times 10^{13} \, H_3O^+ \, ions$$

f. *How many OH^- ions are present?*

$$2.2 \times 10^{-10} \, \cancel{mol} \; \times \; \frac{6.02 \times 10^{23} \, ions}{1 \, \cancel{mol}} \; = \; 1.3 \times 10^{14} \, OH^- \, ions$$

7.67 *Write the chemical equation for the reaction of each weak acid with water. Write the corresponding acidity constant expression.*

Recall that acids donate H^+. Remove one H^+ from the acid given and write the formula remaining, but lower the charge by 1. Make a hydronium ion, H_3O^+, out of the water molecule by adding the H^+ to it. Follow the rules covered previously in Problems 9.15 and 9.19 to write the equilibrium expression.

a. NH_4^+

$$NH_4^+(aq) \; + \; H_2O(l) \; \rightleftharpoons \; H_3O^+(aq) \; + \; NH_3(aq)$$

$$K_a \; = \; \frac{[H_3O^+][NH_3]}{[NH_4^+]}$$

b. HPO_4^{-2}

$$HPO_4^{2-}(aq) \; + \; H_2O(l) \; \rightleftharpoons \; H_3O^+(aq) \; + \; PO_4^{3-}(aq)$$

$$K_a \; = \; \frac{[H_3O^+][PO_4^{3-}]}{[HPO_4^{2-}]}$$

7.69   a. The K_a of NH_4^+ equals 5.6×10^{-10}. For the reaction in Problem 7.67a, what are there more of at equilibrium, reactants or products?

Reactants. The K_a is small.

b. The K_a of HPO_4^{2-} equals 4.2×10^{-13}. For the reaction in Problem 7.67b, what are there more of at equilibrium, reactants or products?

Reactants. The K_a is small.

c. Which is the stronger acid, NH_4^+ or HPO_4^{2-}?

NH_4^+. The K_a of NH_4^+ is greater than the K_a of HPO_4^{2-}.

7.71   *Calculate the pK_a of each acid and indicate which is the stronger acid.*

pK_a is calculated the same way as pH, except that the K_a value is used instead of the $[H_3O^+]$.

a. *HClO, $K_a = 3.0 \times 10^{-8}$*

$pK_a = -\log K_a$

$pK_a = -\log [3.0 \times 10^{-8}]$ On a standard scientific calculator, enter the number 3.0×10^{-8} into the calculator, then press the **log** button. This gives -7.52. Then, - (-7.52) = 7.52. Report the value to two digits to the right of the decimal point.

$pK_a = 7.52$

b. *$C_2O_4H^-$, $K_a = 6.4 \times 10^{-5}$*

$pK_a = -\log K_a$

$pK_a = -\log [6.4 \times 10^{-5}] = 4.19$

$C_2O_4H^-$ is the stronger acid because acid strength increases as pK_a decreases.

7.73   *Write the formula of the conjugate base of each acid in Problem 7.71. Which of the two is the stronger base?*

a. For the acid HClO, ClO⁻ is its conjugate base.

b. For the acid $HC_2O_4^-$, $C_2O_4^{2-}$ is its conjugate base.

Because HClO is the weaker acid, its conjugate base (ClO⁻) is the stronger base.

7.75   *0.10 M solutions of each of the following acids are prepared: acetic acid (K_a = 1.8 x 10⁻⁵) and hydrofluoric acid (K_a = 6.6 x 10⁻⁴). Which acid solution will have the lower pH? Explain.*

Hydrofluoric acid. Hydrofluoric acid has a K_a of 6.6 x 10⁻⁴ while acetic acid has a K_a of 1.8 x 10⁻⁵. The higher the K_a value, the stronger the acid is, and the more acidic the solution (lower pH) is.

7.77   *a. How many moles of NaOH are present in 31.7 mL of 0.155 M NaOH?*

Convert the volume of the NaOH to liters then use the molarity of the NaOH to convert to moles of NaOH.

$$31.7 \; \cancel{mL} \; \times \; \frac{10^{-3} \; \cancel{L}}{1 \; \cancel{mL}} \; \times \; \frac{0.155 \; mol}{1 \; \cancel{L}} \; = \; 4.91 \times 10^{-3} \; mol \; NaOH$$

b. How many moles of HCl are present in a 15.0 mL sample that is neutralized by the 31.7 mL of 0.155 M NaOH?

The reaction for the neutralization is: HCl + NaOH → H_2O + NaCl. The number of moles of HCl reacted should be equal to the number of moles of NaOH present in the base solution:

$$4.91 \times 10^{-3} \; \cancel{mol \; NaOH} \; \times \; \frac{1 \; mol \; HCl}{1 \; \cancel{mol \; NaOH}} \; = \; 4.91 \times 10^{-3} \; mol \; HCl$$

c. What is the molar concentration of the HCl solution described in part b of this question?

The molar concentration is the number of moles of the HCl divided by the volume of the HCl solution given in liters. First, convert the volume of the HCl solution from mL to L and then divide the number of moles from part b by this number:

$$4.91 \times 10^{-3} \text{ mol HCl} \div \left(15.0 \text{ mL} \times \frac{10^{-3} \text{ L}}{\text{mL}} \right) = 0.327 \text{ mol/L or } 0.327 \text{ M}$$

7.79   *It requires 17.2 mL of 0.100 M KOH to titrate 75.0 mL of an HCl solution of unknown concentration. Calculate the initial HCl concentration.*

Titration is a technique used to determine the concentration of an unknown solution. In an acid-base titration, the determination of the concentration is made using the balanced equation for the neutralization reaction between an acid and a base.

In this problem, the concentration of HCl can be determined by calculating the number of moles of HCl that react with the given amount of KOH. The balanced equation is given by:

$$KOH(aq) + HCl(aq) \rightarrow KCl(aq) + H_2O(l)$$

First, determine the number of moles of KOH that completely react with all of the HCl in the solution. Since we are working with molarities, convert 17.2 mL to L first.

$$17.2 \text{ mL} \times \frac{10^{-3} \text{ L}}{1 \text{ mL}} = 0.0172 \text{ L}$$

$$\text{number of moles of KOH} = 0.0172 \text{ L} \times \frac{0.100 \text{ mol KOH}}{1 \text{ L}} = 1.72 \times 10^{-3} \text{ mol KOH}$$

Use the mole ratio from the balanced equation to determine the number of moles of HCl:

$$\text{number of moles of HCl} = 1.72 \times 10^{-3} \text{ mol KOH} \times \frac{1 \text{ mol HCl}}{1 \text{ mol KOH}} = 1.72 \times 10^{-3} \text{ mol HCl}$$

Then, calculate the concentration of HCl. The total volume of the original HCl solution is 750 mL or 0.0750 L.

$$\text{concentration of HCl} = \frac{1.72 \times 10^{-3} \text{ mol HCl}}{0.0750 \text{ L}} = 0.0229 \text{ M HCl}$$

7.81 *It requires 25.9 mL of 0.100 M HCl to titrate 15.0 mL of an NaOH solution of unknown concentration. What is the NaOH concentration?*

In this problem, the concentration of NaOH can be determined by calculating the number of moles of HCl that react with the given volume of NaOH. The balanced equation is given by:

NaOH*(aq)* + HCl*(aq)* → NaCl*(aq)* + H_2O*(l)*

Calculate the number of moles of HCl:

$$25.9 \text{ mL} \times \frac{10^{-3} \text{ L}}{1 \text{ mL}} = 0.0259 \text{ L}$$

$$\text{number of moles of HCl} = 0.0259 \text{ L} \times \frac{0.100 \text{ mol HCl}}{1 \text{ L}} = 2.59 \times 10^{-3} \text{ mol HCl}$$

Use the mole ratio from the balanced equation to determine the number of moles of NaOH:

$$\text{number of moles of NaOH} = 2.59 \times 10^{-3} \text{ mol HCl} \times \frac{1 \text{ mol NaOH}}{1 \text{ mol HCl}} = 2.59 \times 10^{-3} \text{ mol NaOH}$$

Calculate the concentration of NaOH. The total volume of the original NaOH solution is 15.0 mL or 0.0150 L.

$$\text{concentration of HCl} = \frac{2.59 \times 10^{-3} \text{ mol NaOH}}{0.0150 \text{ L}} = 0.173 \text{ M NaOH}$$

7.83 *Ammonium ion (NH_4^+, $pK_a = 9.25$) reacts with water to produce ammonia (NH_3) and hydronium ion (H_3O^+).*

$$NH_4^+ \quad + \quad H_2O \quad \rightleftharpoons \quad NH_3 \quad + \quad H_3O^+$$

a. At pH = 9.25, which of the following is true?
$[NH_4^+] = [NH_3]$, $[NH_4^+] > [NH_3]$, $[NH_4^+] < [NH_3]$.

$[NH_4^+] = [NH_3]$
When the pH of the solution has the same value as the pK_a, the concentration of the acid ($[NH_4^+]$) equals the concentration of its conjugate base ($[NH_3]$).

b. At pH = 5.25, which of the following is true?
$[NH_4^+]$ = $[NH_3]$, $[NH_4^+]$ > $[NH_3]$, $[NH_4^+]$ < $[NH_3]$.

$[NH_4^+]$ > $[NH_3]$
When the pH is lower than the pK_a (5.25 < 9.25), the concentration of the acid is greater than that of the conjugate base.

c. At pH = 13.25, which of the following is true?
$[NH_4^+]$ = $[NH_3]$, $[NH_4^+]$ > $[NH_3]$, $[NH_4^+]$ < $[NH_3]$.

$[NH_4^+]$ < $[NH_3]$
When the pH is greater than the pK_a (13.25 > 9.25), the concentration of the acid is *less* than that of the conjugate base.

7.85   *a. The pK_a of NH_4^+ equals 9.25. For the reaction in Problem 7.67a, what is there more of at pH 7.0, NH_4^+ or NH_3?*

NH_4^+
The pH is less than the pK_a (7.0 < 9.25) therefore, the concentration of the acid ($[NH_4^+]$) is greater than that of the conjugate base ($[NH_3]$).

b. The pK_a of HPO_4^{2-} equals 7.21. For the reaction in Problem 7.67b, what is there more of at pH 10.0, HPO_4^{2-} or PO_4^{3-}?

PO_4^{3-}
The pH is greater than the pK_a (10.0 > 7.21) therefore, the concentration of the acid ($[HPO_4^{2-}]$) is *less* than that of the conjugate base ($[PO_4^{3-}]$).

7.87   *At low pH, the amino acid aspartic acid has the structure shown below.*

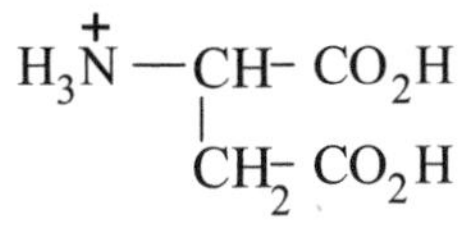

$$H_3\overset{+}{N}-CH-CO_2H$$
$$CH_2-CO_2H$$

Aspartic acid

a. Redraw aspartic acid, showing the acidic- CO_2H and $-NH_3^+$ functional groups in their conjugate base form.

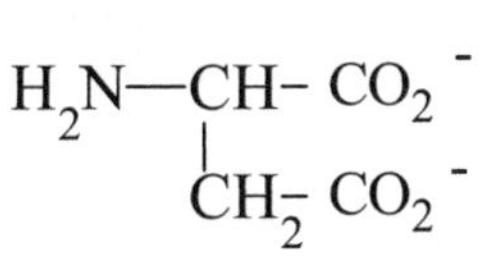

$$H_2N-CH-CO_2^-$$
$$CH_2-CO_2^-$$

b. In amino acids, -CO₂H groups have pK_a's in the range 1.8 – 4.3 and –NH₃⁺ groups have pK_a's in the range 9.1 – 12.5. Draw aspartic acid as it would appear at pH 7. What is the net charge on aspartic acid at this pH?

$$\overset{+}{H_3}N - CH - CO_2^{\ -}$$
$$\mid$$
$$CH_2 - CO_2^{\ -}$$

The net charge is -1.

pH 7 is greater than the pK_a range for -CO_2H groups so the conjugate base form predominates (CO_2^-).
pH 7 is less than the pK_a range for –NH_3^+ groups so the acid form predominates (–NH_3^+).
The net charge is – 1 because +1 -1 -1 = -1.

7.89   *An equilibrium mixture of H₂CO₃ and HCO₃⁻ has a pH of 6.5. What happens to the H₂CO₃ and HCO₃⁻ concentrations when the pH of the solution is adjusted to 8.5?*

First, let's write the relevant equilibrium reaction:

$$H_2CO_3(aq) + H_2O(l) \rightleftharpoons HCO_3^-(aq) + H_3O^+(aq)$$

Initially, the pH of the solution is 6.5. Raising the pH to 8.5 means that the $[H_3O^+]$ will decrease. Decreasing the $[H_3O^+]$ will decrease the rate of the reverse reaction. This will consume HCO_3^- less quickly and will result in an increase in the HCO_3^- concentration. The H_2CO_3 concentration will decrease.

7.91   *A buffer consisting of a weak acid and its conjugate base is prepared. Over what pH range will the buffer be most effective?*

pH = pKa ± 1

7.93   *If you wish to maintain a pH of 6.0, which is the better buffer, H₂CO₃ and HCO₃⁻ or H₂PO₄⁻ and HPO₄²⁻?*

H_2CO_3 and HCO_3^-
Buffers are most resistant to pH changes when the pH equals the pK_a of the weak acid. The pK_a of H_2CO_3 is 6.36 while the pK_a of $H_2PO_4^-$ is 7.21. Because the pK_a of H_2CO_3 is closer to 6.0, this acid and its conjugate base would make a better buffer for a pH of 6.0.

7.95   *A buffer can be prepared using NH_4^+ and NH_3.*
       a. Write an equation for the reaction that takes place when H_3O^+ is added to this buffer.
       b. Write an equation for the reaction that takes place when OH^- is added to this buffer.

A buffer is a solution that resists changes in pH. Buffers commonly contain a weak acid and its conjugate base. The weak acid will neutralize any added base while the conjugate base will neutralize any added acid.

a. Write an equation for the reaction that takes place when H_3O^+ is added to this buffer.

$$NH_3 \quad + \quad H_3O^+ \quad \rightarrow \quad NH_4^+ + H_2O$$

b. Write an equation for the reaction that takes place when OH^- is added to this buffer.

$$NH_4^+ \quad + \quad OH^- \quad \rightarrow \quad NH_3 \quad + \quad H_2O$$

7.97   *Explain how the H_2CO_3/HCO_3^- buffer system helps maintain blood serum at a constant pH.*

This buffer makes use of the equilibrium

$$H_2CO_3 + H_2O \rightleftharpoons HCO_3^- + H_3O^+$$

In response to an increase in $[H_3O^+]$ (a drop in pH) the equilibrium shifts to the left.

In response to a decrease in $[H_3O^+]$ (a rise in pH) the equilibrium shifts to the right.

7.99   *Explain how respiration helps maintain blood serum at a constant pH.*

In response to a drop in blood pH (a rise in $[H_3O^+]$) the rate of respiration increases. As more CO_2 is exhaled, $[H_2CO_3]$ drops, as does $[H_3O^+]$.

$$H_2O + CO_2 \leftarrow H_2CO_3$$

$$H_2CO_3 + H_2O \leftarrow HCO_3^- + H_3O^+$$

In response to a rise in blood pH (a drop in $[H_3O^+]$) the rate of respiration slows. As CO_2 accumulates, $[H_2CO_3]$ increases, as does $[H_3O^+]$.

$$H_2O + CO_2 \rightarrow H_2CO_3$$

$$H_2CO_3 + H_2O \rightarrow HCO_3^- + H_3O^+$$

7.101 *Explain how the kidneys help maintain blood serum at a constant pH.*

When the pH of blood serum becomes too low (high H_3O^+), the kidneys release HCO_3^- ions that neutralize excess H_3O^+. When the pH of the blood serum becomes too high, the kidneys do the opposite: HCO_3^- is removed from the blood causing the concentration of H_3O^+ to increase, thereby lowering the pH.

7.103 *Explain how hyperventilation can lead to respiratory alkalosis.*

Hyperventilation causes CO_2 to be exhaled faster than it can be formed by the cells. The decrease in CO_2 causes a decrease in H_2CO_3. The pH increases resulting in alkalosis.

7.105 *According to Section 7.11, how does cystic fibrosis upset the blood buffer system to cause acidosis?*

Cystic fibrosis is one of many conditions that hamper the ability of the lungs to exchange gases. Insufficient exchange of gases in the lungs can lead to a build-up of CO_2. Increased CO_2 concentration leads to a higher concentration of H_2CO_3. This shifts the blood buffer system reaction,

$$H_2CO_3 + H_2O \rightleftharpoons HCO_3^- + H_3O^+$$

to produce more H_3O^+, which results in a lower pH.

7.107 *In terms of Le Chatelier's principle, describe how the equilibrium $Mb + O_2 \rightleftharpoons MbO_2$ responds to increases in the concentration of*

a. Mb

When [Mb] is increased, the rate of the forward reaction increases and more MbO_2 is produced.

b. O₂

When [O_2] is increased, the rate of the forward reaction increases and more MbO_2 is produced.

c. MbO₂

When [MbO_2] is increased, the rate of the reverse reaction increases and more Mb and O_2 is produced.

7.109 *Calculate the logarithm of each number. Assume that each is a measured quantity.*

a. 1 x 10⁻⁹

Using a scientific calculator, enter the number 1×10^{-9}, then press the **log** button. This gives -9. Report the log with one digit to the right of the decimal point.

-9.0

b. 1 x 10¹²

Using a scientific calculator, enter the number 1×10^{12}, then press the **log** button. This gives 12. Report the log with one digit to the right of the decimal point.

12.0

c. 3.4 x 10⁻²

Using a scientific calculator, enter the number 3.4×10^{-2}, then press the **log** button. This gives -1.468. Report the log with two digits to the right of the decimal point.

-1.47

d. 9.7 x 10⁴

Using a scientific calculator, enter the number 9.7×10^{4}, then press the **log** button. This gives 4.98677. Report the log with two digits to the right of the decimal point.

4.99

7.111   *Calculate the antilogarithm of each number.*

The numbers given in this problem are not said to be measured quantities, so the number of digits reported in the answers is arbitrary.

a. 8

Using a scientific calculator, enter the number 8, then press the **2nd** button followed by the **log** button. This gives 1×10^8.

1×10^8

b. –3

Using a scientific calculator, enter the number -3 (remember to put in 3 then press the sign change, NOT the subtraction button) then press the **2nd** button followed by the **log** button. This gives 1×10^{-3}.

1×10^{-3}

c. 7.9

Using a scientific calculator, enter the number 7.9 into the calculator, then press the **2nd** button followed by the **log** button. This gives 7.943×10^7.

7.9×10^7

d. 15.3

Using a scientific calculator, enter the number 15.3, then press the **2nd** button followed by the **log** button. This gives 1.995×10^{15}.

2.0×10^{15}

7.113   *Someone tells you that the spice turmeric is a natural pH indicator. How might you test this claim, using only substances that you are likely to have available at home?*

Test glass cleaner (containing basic ammonia) or soap and the pH should indicate a basic solution. Test vinegar and the pH should indicate an acidic solution.

7.115 *a. Write the balanced chemical equation for the reversible acid-base reaction that takes place between nitrous acid (HNO_2) and water.*

$$HNO_2 \quad + \quad H_2O \quad \leftrightharpoons \quad NO_2^- \quad + \quad H_3O^+$$

b. Write the equilibrium constant expression for this reaction.

$$K_a = \frac{[NO_2^-][H_3O^+]}{[HNO_2]}$$

c. Name and draw the line-bond structure of the conjugate base of nitrous acid.

The conjugate base of nitrous acid (HNO_2) is the nitrite ion (NO_2^-). Its line-bond structure is:

$$\left[:\ddot{O}=\ddot{N}-\ddot{O}: \right]^-$$

d. For nitrous acid, $K_a = 4.0 \times 10^{-4}$. What is the value of pK_a?

$pK_a = -\log K_a = -\log 4.0 \times 10^{-4} = 3.40$

e. Which are there more of once the reaction in part a has reached equilibrium, reactants or products?

Reactants. The value of the K_a is small.

f. For carbonic acid (H_2CO_3), $K_a = 4.4 \times 10^{-7}$. Which is the stronger acid, nitrous acid or carbonic acid?

HNO_2 or nitrous acid. The K_a of nitrous acid (4.0×10^{-4}) is greater than the K_a of carbonic acid 4.4×10^{-7}.

g. Which is the stronger base, the conjugate base of nitrous acid or the conjugate base of carbonic acid?

Conjugate base of carbonic acid. Carbonic acid is a weaker acid than nitrous acid.

h. Assuming that the reaction in your answer to part a is at equilibrium, what will happen to restore equilibrium if the pH is lowered by addition of H_3O^+, a net forward reaction or a net reverse reaction?

Net reverse reaction. A net reverse reaction will occur to consume the additional H_3O^+ and restore the equilibrium ratio of product concentrations to reactant concentrations.

i. Over what pH range would a solution containing nitrous acid and its conjugate base be effective as a buffer?

pH = $K_a \pm 1$ = 3.40 ± 1

7.117 *Suggest an explanation for the fact that the antacids Di-Gel, Mylanta, and Maalox contain both Al(OH)₃ and Mg(OH)₂.*

The constipating effect of Al^{3+} is offset by the laxative effect of Mg^{2+}.

Chapter 8
Carboxylic Acids, Phenols, and Amines

Solutions to Problems

8.1 *The ester methyl salicylate (wintergreen oil) can be prepared as shown below. Circle the atoms in the reactant molecules that are the source of the water molecule that appears in the product.*

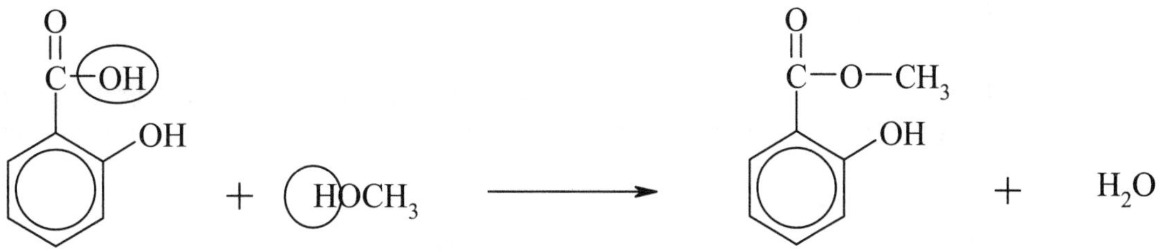

8.3 *Name each carboxylic acid.*

Follow the IUPAC rules for naming carboxylic acids. The parent chain is enclosed in a box and any substituent group is circled. Begin numbering the carbon atoms on the parent chain starting from the C atom of the carboxyl group.

a.   3,5-dimethylhexanoic acid

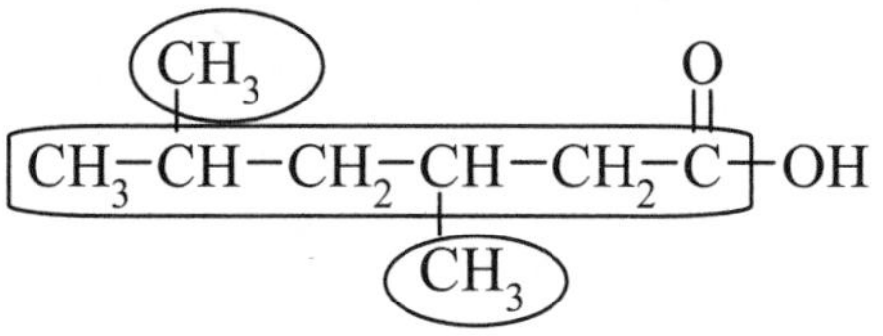

b.   propanoic acid

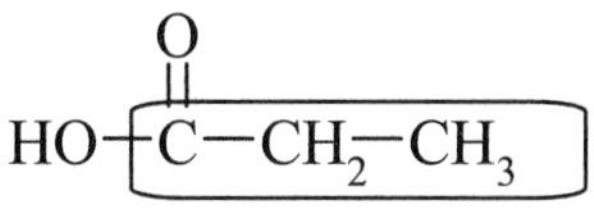

c.   2-propyldecanoic acid

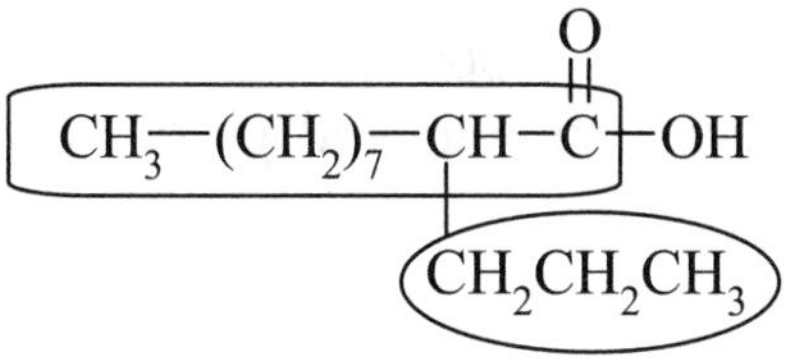

8.5   *Draw each carboxylic acid.*

a. octanoic acid

Start by drawing CH_3 then add $(CH_2)_6$ followed by the ending carboxyl group.

$$CH_3-(CH_2)_6-\overset{\overset{\displaystyle O}{\|}}{C}-OH$$

b. 3,3-dimethylheptanoic acid

First, draw the parent chain with seven (heptane) carbons making the last one the carboxyl group indicated by the "oic acid" ending. Add 2 methyl groups to carbon 3.

$$CH_3-CH_2-CH_2-CH_2-\overset{\overset{\displaystyle CH_3}{|}}{\underset{\underset{\displaystyle CH_3}{|}}{C}}-CH_2-\overset{\overset{\displaystyle O}{\|}}{C}-OH$$

c. 3-isopropylhexanoic acid

First, draw the parent chain with six (hexane) carbons making the last one the carboxyl group indicated by the "oic acid" ending. Add an isopropyl group to carbon 3.

$$CH_3-CH_2-CH_2-\underset{\underset{\displaystyle CH_3CHCH_3}{|}}{CH}-CH_2-\overset{\overset{\displaystyle O}{\|}}{C}-OH$$

d. 2-bromopentanoic acid

First, draw the parent chain with five (pentane) carbons making the last one the carboxyl group indicated by the "oic acid" ending. Add a bromine atom to carbon 2.

$$CH_3-CH_2-CH_2-\underset{\underset{\displaystyle Br}{|}}{CH}-\overset{\overset{\displaystyle O}{\|}}{C}-OH$$

8.7   *Match each structure to the correct IUPAC name: 3-methylbutanoic acid, 2-methylpentanoic acid, 2-methylbutanoic acid.*

The carboxylic functional group is dominant in the naming so its carbon is carbon 1. In drawing a, there are four carbons and the methyl group is on the second carbon so that the compound is 2-methylbutanoic acid. Drawing b also has four carbons but the methyl group is on carbon 3, making its name 3-methylbutanoic acid. In drawing c, the methyl is on the second carbon but the parent chain has five carbons which makes the name 2-methylpentanoic acid.

a.

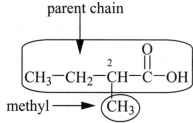

parent chain

$CH_3-CH_2-\underset{2}{CH}-\overset{\displaystyle O}{\overset{\|}{C}}-OH$

methyl ⟶ CH_3

2-methylbutanoic acid

b.

parent chain

$CH_3-\underset{3}{CH}-CH_2-\overset{\displaystyle O}{\overset{\|}{C}}-OH$

methyl ⟶ CH_3

3-methylbutanoic acid

c.

methyl ⟶ $H_3C)-\underset{2}{CH}-\overset{\displaystyle O}{\overset{\|}{C}}-OH$ ⟵ parent chain

$H_2C-CH_2-CH_3$

2-methylpentanoic acid

190

8.9 *Acetic acid (CH$_3$CO$_2$H) and propyl alcohol (CH$_3$CH$_2$CH$_2$OH) have the same molecular weight, but their boiling points differ by about 20°C (acetic acid, 118°C; propyl alcohol, 97.4°C). Account for this difference.*

Acetic acid molecules form more hydrogen bonds with one another than do propyl alcohol molecules. Intermolecular bonds cause a compound to have a high boiling point.

8.11 *Name each phenol.*

Follow the IUPAC rules for naming phenol compounds. Begin numbering the C atoms in the ring starting at the C atom to which the –OH group is attached. Count in the direction (clockwise or counterclockwise) that assigns the lower number to the nearest substituent. For disubstituted rings, the prefixes *ortho-*, *meta-*, or *para-* may be used to indicate the position of a substituent group relative to the –OH group. In some cases, common names are also used.

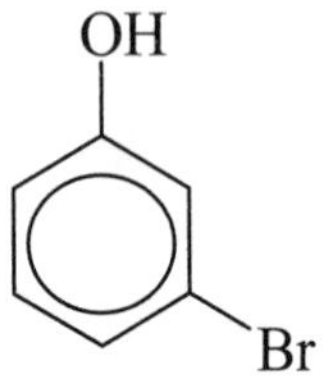

a.   3-bromophenol or m-bromophenol

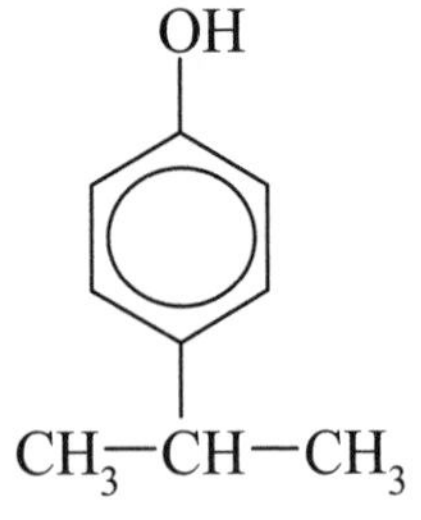

b.   4-isopropylphenol

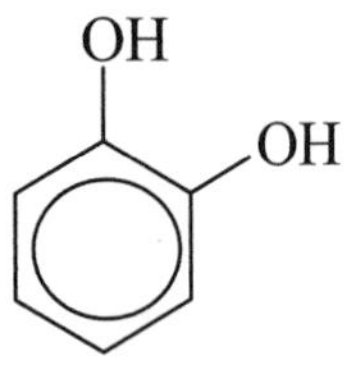

c.   catechol

191

8.13   *Draw each phenol.*

a. 2,4-dimethylphenol

Start by drawing the phenol parent (benzene ring with a –OH in the first position). Counting from the –OH carbon add one methyl group to carbon 2 and one methyl group to carbon 4.

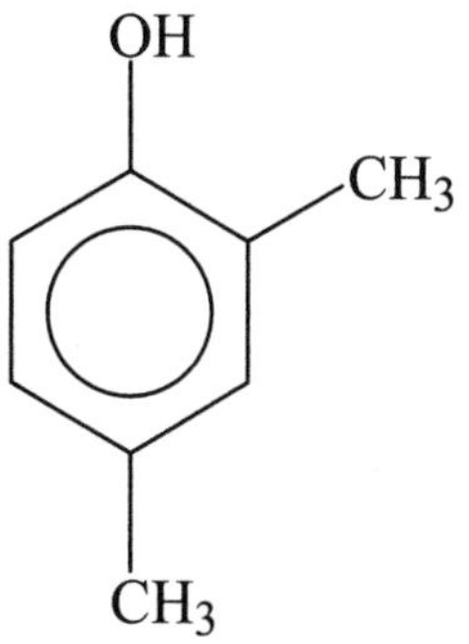

b. 3,5-dimethylphenol

Start by drawing the phenol parent (benzene ring with a –OH in the first position). Counting from the –OH carbon add one methyl group to carbon 3 and one methyl group to carbon 5.

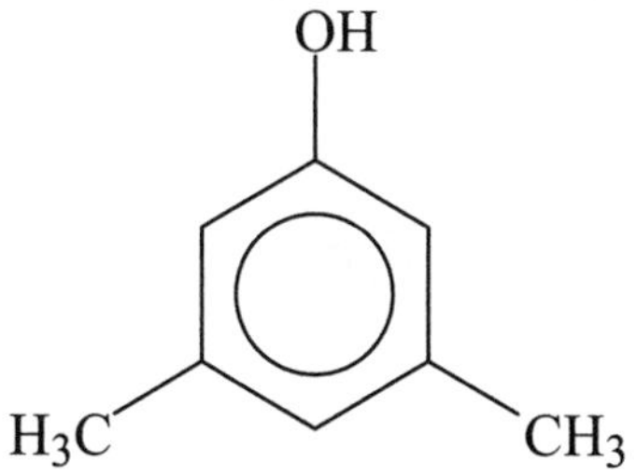

c. 4-propylphenol

Start by drawing the phenol parent (benzene ring with a –OH in the first position). Counting from the –OH carbon add a propyl group to carbon 4 (a propyl group is propane attached by its carbon 1).

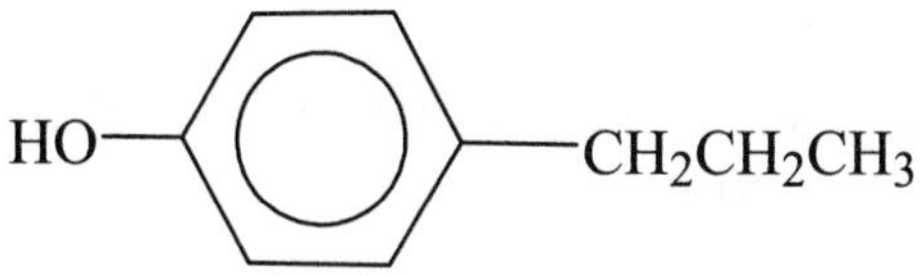

d. m-*propylphenol*

Start by drawing the phenol parent (benzene ring with a –OH in the first position). Counting from the –OH carbon add a propyl group to carbon 3 so the two substituents are meta to each other.

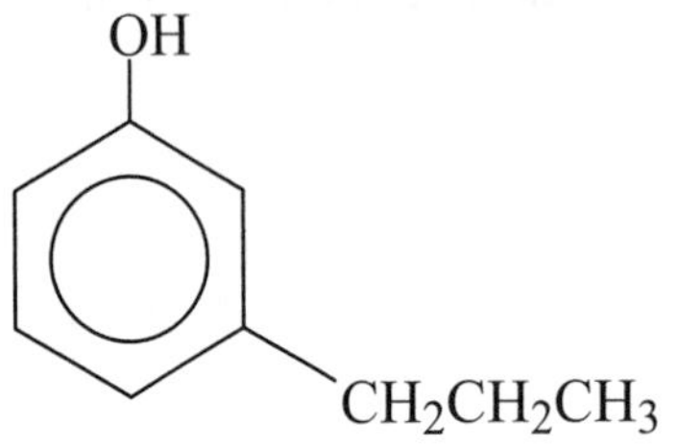

8.15   *Hydroquinone (Table 8.2) has a higher boiling point than phenol. What might account for this difference?*

Hydroquinone molecules, with two –OH groups, can form more hydrogen bonds with one another than do phenol molecules.

8.17   *Is malvidin (Figure 8.1b) a catechol-, a resorcinol-, or a hydroquinone-containing compound?*

It is a resorcinol-containing compound because it has two –OH groups in the *meta* configuration in an aromatic ring.

8.19   *Erbstatin has shown antitumor activity in combating skin, breast, and esophageal cancer.*

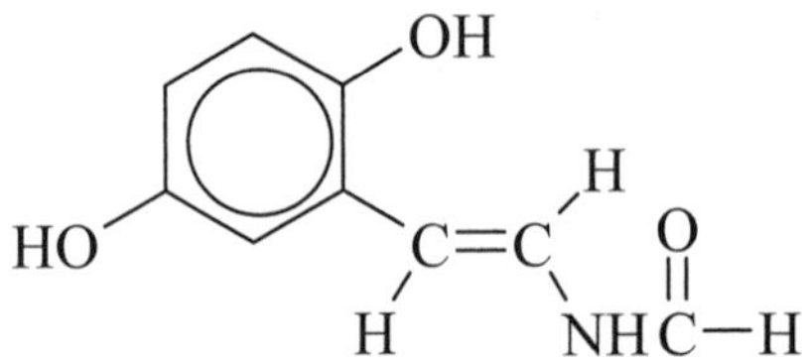

Erbstatin

a. Is erbstatin a catechol-, a resorcinol-, or a hydroquinone-containing phenol?

Erbstatin is a hydroquinone-containing phenol because it contains a phenol group with two –OH groups in *para* orientation.

b. Is the nitrogen atom present in this compound part of an amine functional group?

No. The nitrogen atom is not part of an amine functional group because it is bonded to a C=O. It is part of an amide functional group.

c. Which stereoisomer (cis or trans) is present?

The *trans* stereoisomer is present.

8.21 *Hexylresorcinol (4-hexyl-3-hydroxyphenol) is an antiseptic used in some mouthwashes and throat lozenges. Draw this compound.*

Start by drawing the phenol parent (benzene ring with a –OH in the first position). Counting from the –OH carbon add another –OH group (called hydroxy when used as a side chain) to carbon 3. Next add a hexyl to carbon 4.

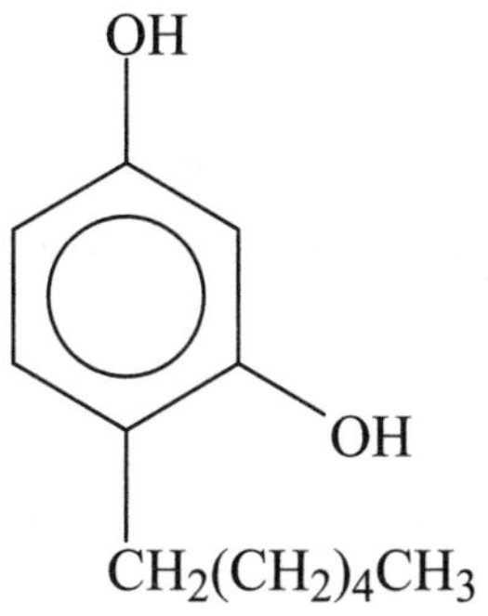

OH

OH

$CH_2(CH_2)_4CH_3$

4-hexyl-3-hydroxyphenol

8.23 *Draw octanoic acid and its conjugate base.*

The conjugate base of an acid is the ion or compound left after a H^+ ion leaves.

$$CH_3-CH_2-CH_2-CH_2-CH_2-CH_2-CH_2-\overset{\overset{O}{\|}}{C}-OH$$

octanoic acid

$$CH_3-CH_2-CH_2-CH_2-CH_2-CH_2-CH_2-\overset{\overset{O}{\|}}{C}-O^-$$

octanoate ion (conjugate base)

8.25 *Draw the conjugate base of each carboxylic acid.*

    a. propanoic acid

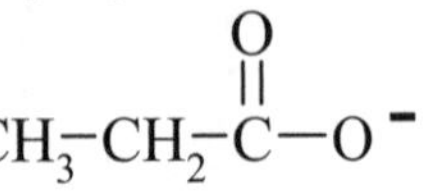

$$CH_3-CH_2-\overset{\overset{\displaystyle O}{\|}}{C}-O^-$$

    b. hexanoic acid

$$CH_3-CH_2-CH_2-CH_2-CH_2-\overset{\overset{\displaystyle O}{\|}}{C}-O^-$$

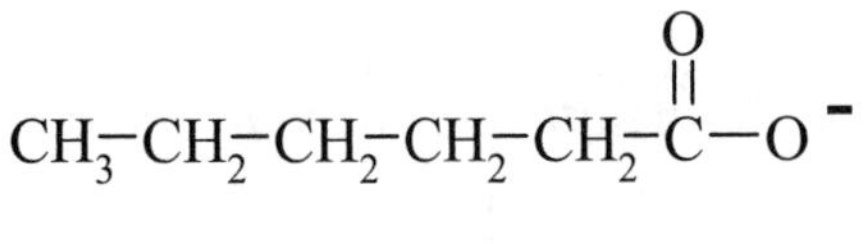

    c. 3-chloropentanoic acid

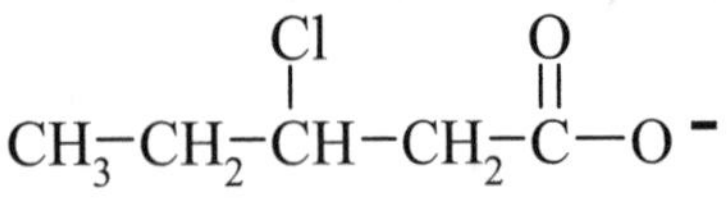

$$CH_3-CH_2-\overset{\overset{\displaystyle Cl}{|}}{C}H-CH_2-\overset{\overset{\displaystyle O}{\|}}{C}-O^-$$

8.27 *Name each of the conjugate bases in your answer to Problem 8.25.*

    a. $CH_3-CH_2-\overset{\overset{\displaystyle O}{\|}}{C}-O^-$ propanoate ion

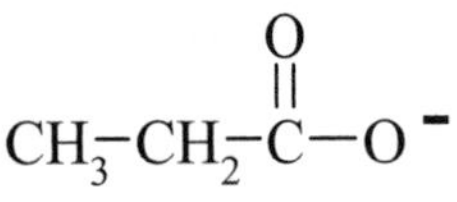

    b. $CH_3-CH_2-CH_2-CH_2-CH_2-\overset{\overset{\displaystyle O}{\|}}{C}-O^-$ hexanoate ion

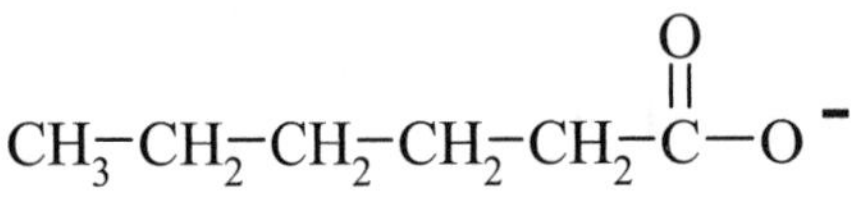

    c. $CH_3-CH_2-\overset{\overset{\displaystyle Cl}{|}}{C}H-CH_2-\overset{\overset{\displaystyle O}{\|}}{C}-O^-$ 3-chloropentanoate ion

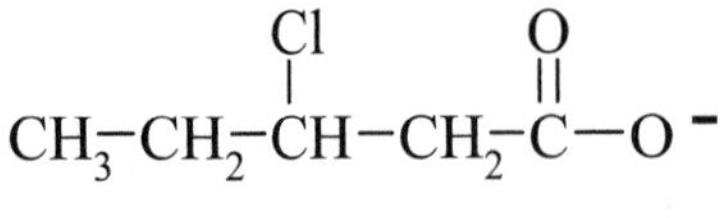

8.29 *Draw and name the conjugate base of the carboxylic acid reactant shown in Figure 8.10a.*

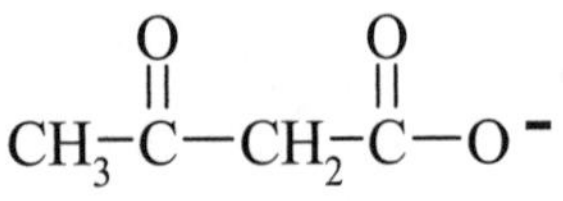

$$CH_3-\overset{\overset{\displaystyle O}{\|}}{C}-CH_2-\overset{\overset{\displaystyle O}{\|}}{C}-O^-$$

    acetoacetate ion

195

8.31 *Draw the products of each reaction.*

In a previous chapter, you saw that acid and base reactions produced a salt and water. The same is true when an organic acid is reacted with a base. The cation of the base forms a salt (ionic compound) with the carboxylate anion formed with the removal of an H^+ from the carboxylic acid.

a.

$$CH_3-CH_2-\underset{\underset{CH_3}{|}}{CH}-\overset{\overset{O}{||}}{C}-OH \ + \ KOH \longrightarrow$$

$$\boxed{CH_3-CH_2-\underset{\underset{CH_3}{|}}{CH}-\overset{\overset{O}{||}}{C}-O^- \ K^+ \ + \ H_2O}$$

b.

$$CH_3-\bigcirc-CH_2CH_2\overset{\overset{O}{||}}{C}-OH \ + \ NaOH \longrightarrow$$

$$\boxed{CH_3-\bigcirc-CH_2CH_2\overset{\overset{O}{||}}{C}-O^- \ Na^+ \ + H_2O}$$

8.33 *Dinoseb, an herbicide and insecticide, is sold as a water-soluble ammonium salt. Draw this ammonium salt, which is produced by reacting Dinoseb with the base ammonia.*

$$CH_3CH_2\underset{\underset{CH_3}{\overset{|}{}}}{CH}-\bigcirc\underset{NO_2}{\overset{OH}{\underset{}{}}}NO_2$$

Dinoseb

The anion of the salt has the original acid structure with the hydrogen removed from the –OH and a negative charge placed on the oxygen. The cation of the salt is drawn beside the O⁻.

8.35   *In 1875 sodium salicylate, the sodium salt of salicylic acid, was introduced as an analgesic (painkiller).*

Salicylic acid

a. *Which functional group of salicylic acid is the most acidic, the phenol group or the carboxyl group?*

The carboxylic acid group is more acidic than the phenol group.

b. *Draw the ionic compound sodium salicylate. (Hint: Only one H⁺ is removed from salicylic acid.)*

8.37   *p*-Ethylphenoxide ion is the conjugate base of *p*-ethylphenol. Draw both compounds.

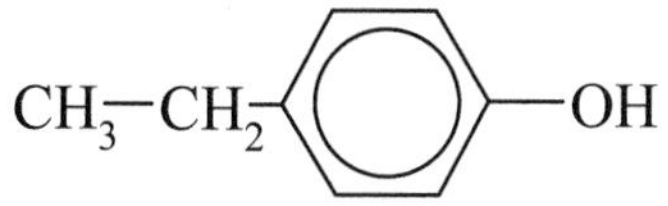

CH$_3$-CH$_2$ [benzene ring] -OH

p-ethylphenol

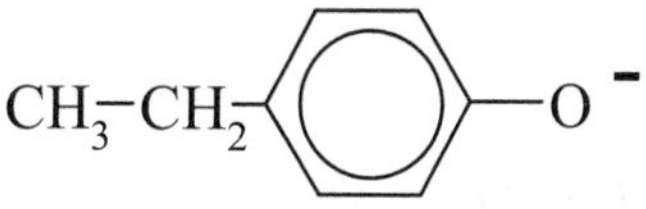

CH$_3$-CH$_2$ [benzene ring] -O $^-$

p-ethylphenoxide

8.39   *a. Draw 4-chloropentanoic acid.*

First, draw pentanoic acid. Starting from carbon 1 (-COOH), count to the 4th carbon and attach a chlorine atom.

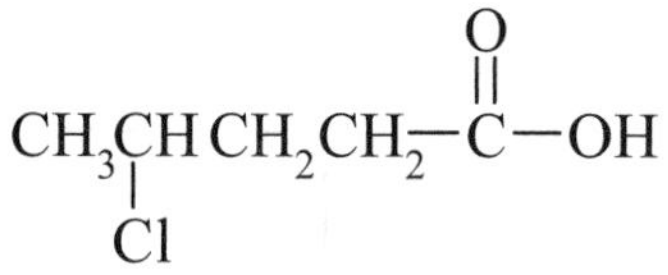

$$CH_3CHCH_2CH_2-\overset{\overset{\displaystyle O}{\displaystyle \|}}{C}-OH$$
$$\overset{|}{Cl}$$

b. Draw and name the conjugate base of 4-chloropentanoic acid.

The conjugate base of an acid has a H$^+$ ion removed and replaced by a negative charge:

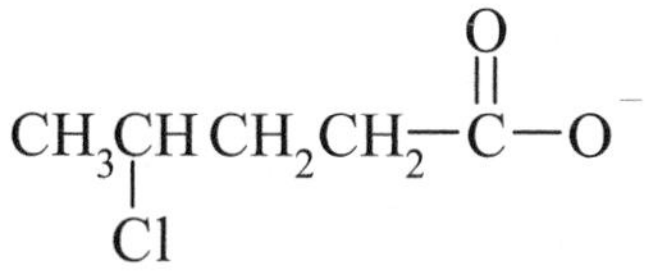

$$CH_3CHCH_2CH_2-\overset{\overset{\displaystyle O}{\displaystyle \|}}{C}-O^-$$
$$\overset{|}{Cl}$$

4-chloropentanoate ion

c. Which predominates in water at pH 7, 4-chloropentanoic acid or its conjugate base?

Its conjugate base. Since most carboxylic acids have pK$_a$ values around 5, carboxylate ions predominate at pH 7.

8.41 *Draw the products of each reaction.*

The reaction of a carboxylic acid and alcohol in the presence of an acid catalyst produces an ester and a water molecule.

a.

b.

8.43 *Name the organic products in Problem 8.41 Hint: -CH$_2$C$_6$H$_5$ is a benzyl group.*

a. propyl benzoate   *b.* benzyl propanoate

8.45 *Draw and name the ester formed when p-ethylbenzoic acid reacts with isopropyl alcohol, $CH_3CH(OH)CH_3$, in the presence of H^+.*

p-ethylbenzoic acid   isopropyl alcohol   isopropyl p-ethylbenzoate

8.47 *Draw the products obtained from each reaction.*

These are hydrolysis reactions in which an ester is broken apart by water to form a carboxylic acid plus an alcohol.

a.

b.

8.49 *Draw the products obtained when*

a. *cyclopentyl acetate reacts with H_2O in the presence of H^+.*

200

b. *cyclopentyl acetate reacts with NaOH(aq).*

In the presence of NaOH, a H^+ ion is removed from the carboxylic acid product and a caboxylate salt is formed.

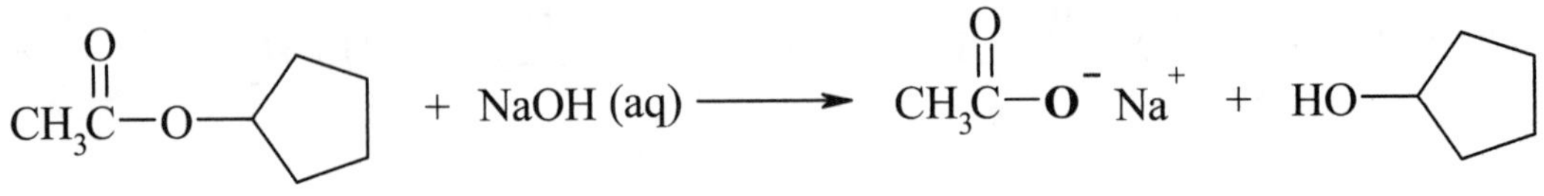

8.51 *Draw the products of each reaction.*

a.

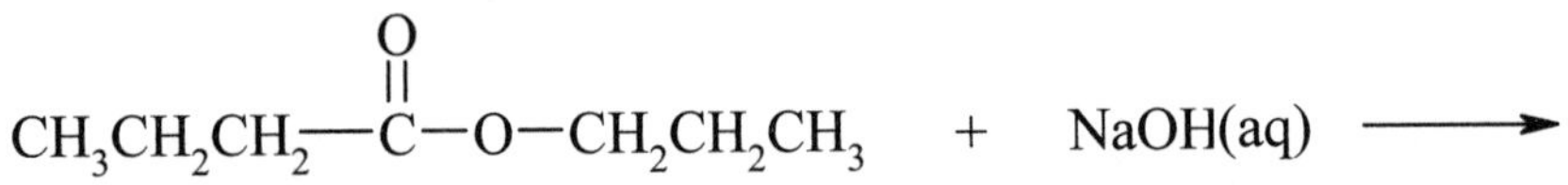

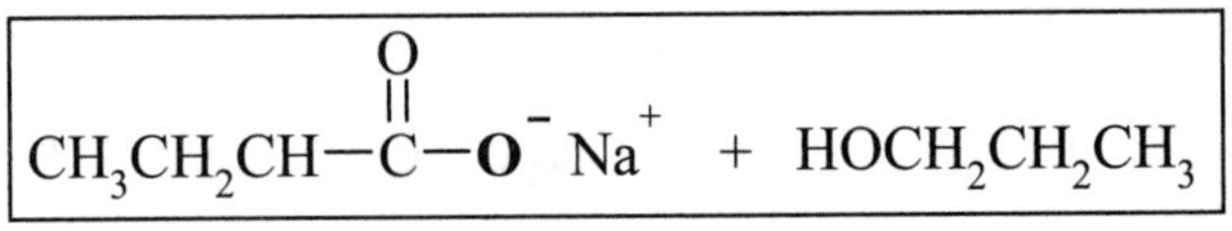

$$CH_3CH_2CH_2-\overset{\overset{\displaystyle O}{||}}{C}-O-CH_2CH_2CH_3 \quad + \quad NaOH(aq) \longrightarrow$$

$$\boxed{CH_3CH_2CH-\overset{\overset{\displaystyle O}{||}}{C}-O^-\ Na^+ \quad + \quad HOCH_2CH_2CH_3}$$

b.

$$HC-O-CH_2CH_3 \quad + \quad KOH\ (aq) \longrightarrow$$

$$\boxed{HC-O^-\ K^+ \quad + \quad HOCH_2CH_3}$$

c.

+ NaOH (aq) ⟶

201

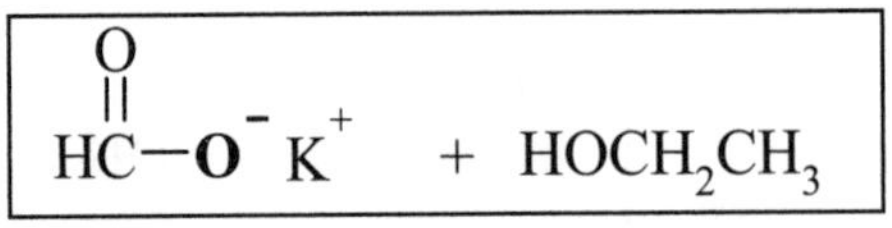

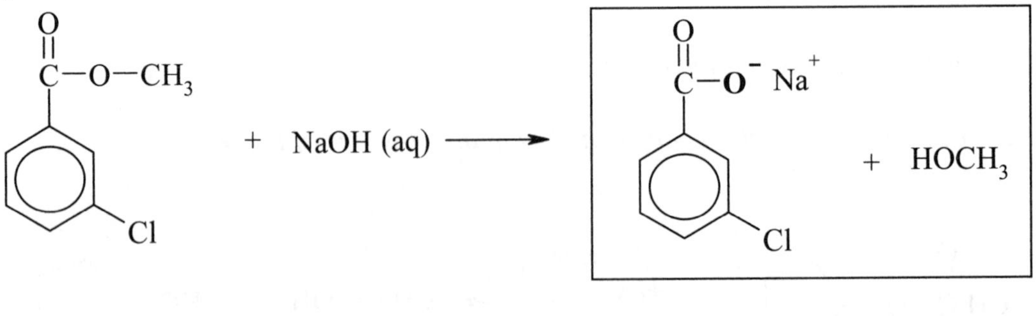

8.53 *Draw a skeletal structure for each of the products formed when the ester below, one of the primary compounds in beeswax, is reacted with NaOH(aq).*

$$CH_3(CH_2)_{14}-\overset{\overset{\displaystyle O}{\|}}{C}-O-CH_2(CH_2)_{28}CH_3$$

The carboxylate ion and alcohol products are:

$+$

8.55 *Draw the products formed when each β-keto acid is decarboxylated.*

When a carboxylic acid is decarboxylated, a carbon dioxide molecule (CO_2) is formed from the carboxyl group (circled).

a.

b. $CH_3CH \overset{\overset{\displaystyle O}{\|}}{C} CH_2 \overset{\overset{\displaystyle O}{\|}}{C}-OH$

8.57 *Draw the products formed when the α-keto acid undergoes an enzyme-catalyzed decarboxylation.*

When a carboxylic acid is decarboxylated, a carbon dioxide molecule (CO_2) is formed from the carboxyl group.

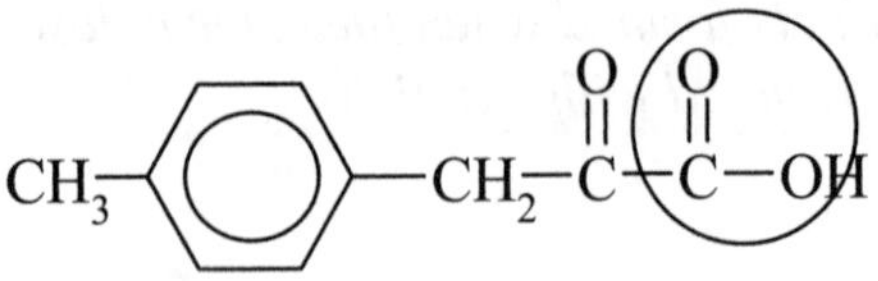

enzyme-catalyzed decarboxylation

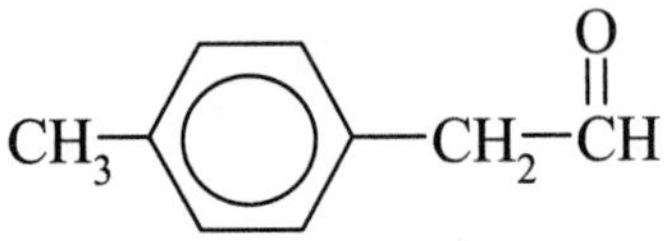

CH_3—⬡—CH_2—$\overset{O}{\underset{\|}{C}}$—CH

8.59   *2-Ethyl-1,4-hydroquinone is an aggregation pheromone (chemical messenger) of the African rhinoceros beetle. Draw the product formed when this compound reacts with $K_2Cr_2O_7$.*

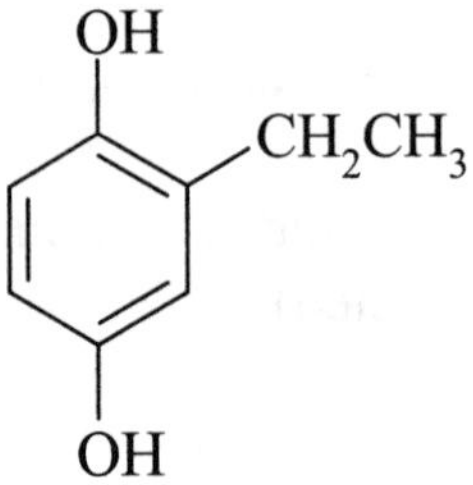

OH

CH_2CH_3

OH

2-Ethyl-1,4-hydroquinone

$K_2Cr_2O_7$ is an oxidizing agent that converts the –OH groups into ketone groups:

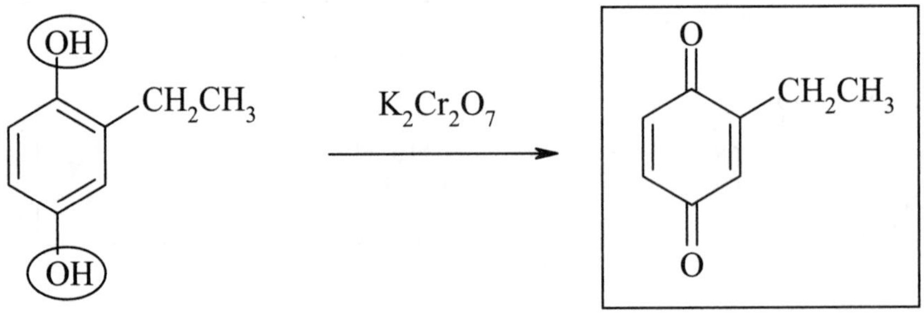

OH

CH_2CH_3

OH

$\xrightarrow{K_2Cr_2O_7}$

O

CH_2CH_3

O

2-ethyl-1,4-hydroquinone

8.61   *Anthraquinone is used in the manufacture of some dyes and is also used by farmers to make seeds distasteful to birds. Draw the hydroquinone-containing compound from which anthraquinone might be produced through oxidation.*

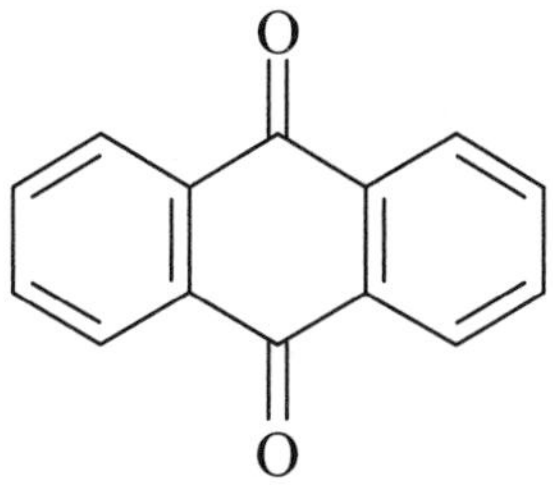

Anthraquinone

The ketone groups in anthraquinone can be formed through oxidation of –OH groups:

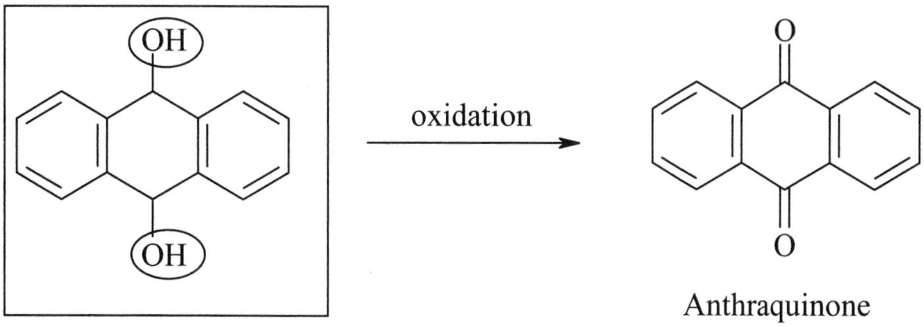

oxidation

Anthraquinone

8.63 *Draw each compound.*

Review the general IUPAC rules for naming amines and alkylammonium ions to derive and draw the structure for each compound or ion below:

a. 1-pentanamine

$CH_3-CH_2-CH_2-CH_2-CH_2-NH_2$

b. N-isopropyl-1-pentanamine

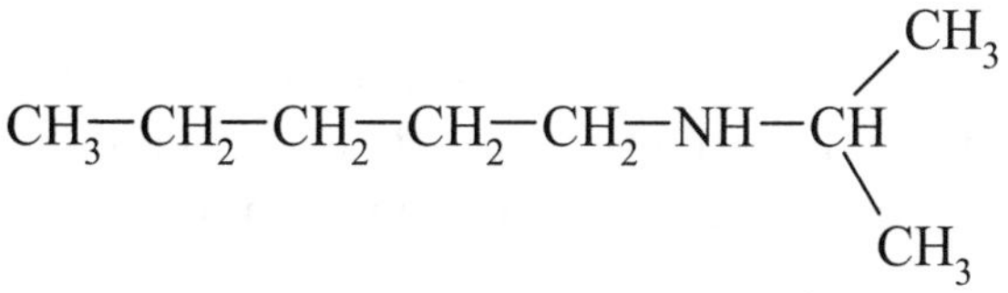

c. N-ethyl-N-methyl-2-hexanamine

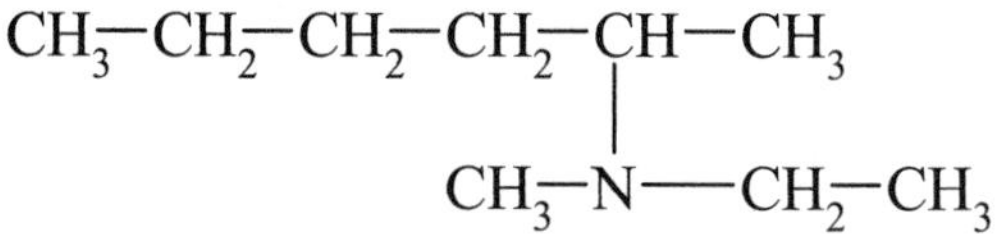

204

d. dimethyldipropylammonium ion

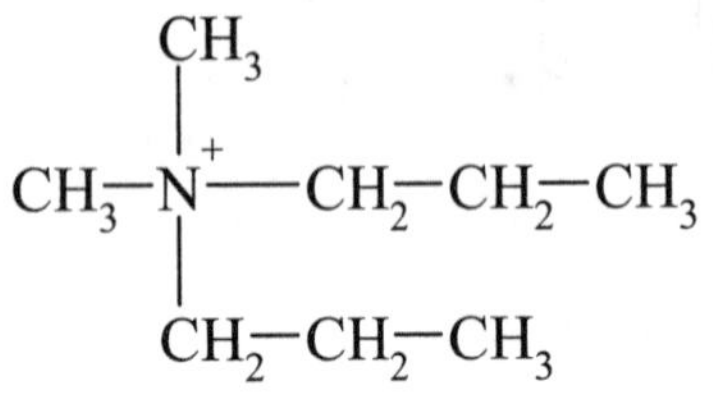

$$CH_3-\overset{\overset{\displaystyle CH_3}{|}}{\underset{\underset{\displaystyle CH_2-CH_2-CH_3}{|}}{N^+}}-CH_2-CH_2-CH_3$$

8.65   *Draw each amine.*

a. ethylamine

$$CH_3-CH_2-NH_2$$

b. isopropylamine

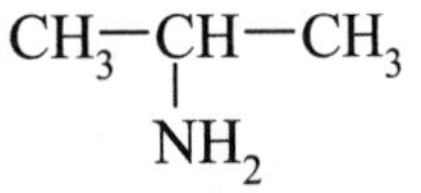

$$CH_3-\overset{}{\underset{\underset{\displaystyle NH_2}{|}}{CH}}-CH_3$$

c. butylmethylamine

$$CH_3-CH_2-CH_2-CH_2-NH-CH_3$$

d. diethylpropylamine

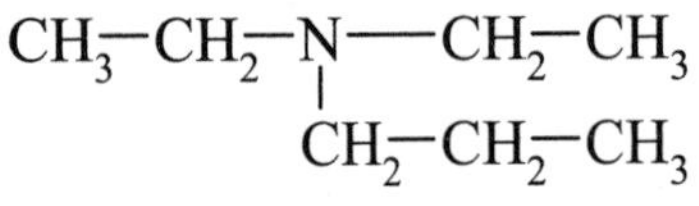

$$CH_3-CH_2-\overset{}{\underset{\underset{\displaystyle CH_2-CH_2-CH_3}{|}}{N}}-CH_2-CH_3$$

8.67   *Give the IUPAC name of each amine.*

The IUPAC naming of amines uses the name of the corresponding alkane parent chain with the "e" replaced with "amine". If a substituent is attached to the nitrogen of the amine, it is preceded by *N–*. The amine position on the parent chain is given by a number preceding the parent chain.

a. CH_3NH_2

methanamine

b. $CH_3CH\,CH_3$
   $\quad\quad\underset{|}{}$
   $\quad\quad NH_2$

2-propanamine

c. $\quad CH_3CH\,CH_3$
   $\quad\quad\underset{|}{}$
   $\quad HN\,CH_2CH_2CH_2CH_3$

N-isopropyl-1-butanamine

d. $CH_3CH_2CH_2\underset{|}{N}\,CH_2CH_3$
   $\quad\quad\quad\quad\quad CH_3$

N-ethyl-*N*-methyl-1-propanamine

8.69   *Give the common name of each amine in Problem 8.67.*

The common name of amines puts the name of all substituents attached to the N in front of "amine".

a. methylamine

b. isopropylamine

c. butylisopropylamine

d. ethylmethylpropylamine

8.71   *Identify each amine in Problem 8.67 as being 1°, 2°, or 3°.*

A 1° amine has only one carbon atom bonded to the amine nitrogen atom. A 2° amine has two carbon atoms each one directly bonded to the amine nitrogen atom. A 3° amine has three carbon atoms each one directly bonded to the amine nitrogen atom.

a. CH$_3$NH$_2$

1° amine

b. 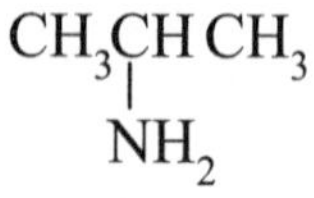 $\underset{\displaystyle NH_2}{CH_3\overset{|}{CH}\,CH_3}$

1° amine

c. 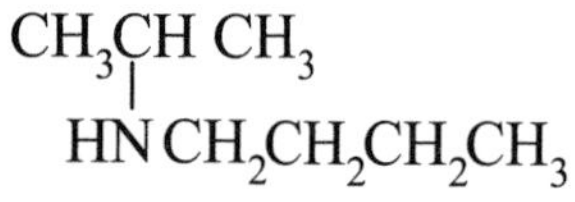 $\underset{\displaystyle HN\,CH_2CH_2CH_2CH_3}{CH_3\overset{|}{CH}\,CH_3}$

2° amine

d. 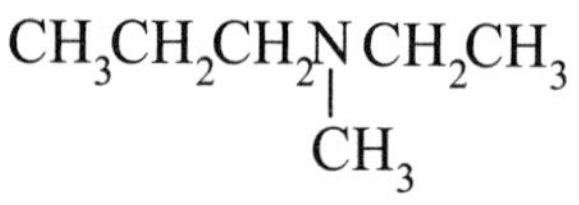 $\underset{\displaystyle CH_3}{CH_3CH_2CH_2\overset{|}{N}\,CH_2CH_3}$

3° amine

8.73 *Identify each compound in Problem 8.63 as being 1°, 2°, 3°, or 4°.*

A 1° amine has only one carbon atom bonded to the amine nitrogen atom. A 2° amine has two carbon atoms, each one directly bonded to the amine nitrogen atom. A 3° amine has three carbon atoms, each one directly bonded to the amine nitrogen atom. A 4° has four carbon atoms, each directly bonded to the amine nitrogen atom.

a. CH$_3$—CH$_2$—CH$_2$—CH$_2$—CH$_2$—NH$_2$ 1° amine

b. 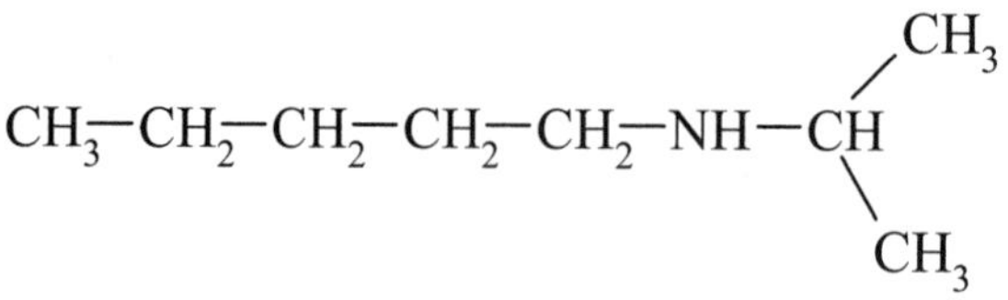 CH$_3$—CH$_2$—CH$_2$—CH$_2$—CH$_2$—NH—CH$\overset{\displaystyle \diagup CH_3}{\diagdown CH_3}$ 2° amine

c. 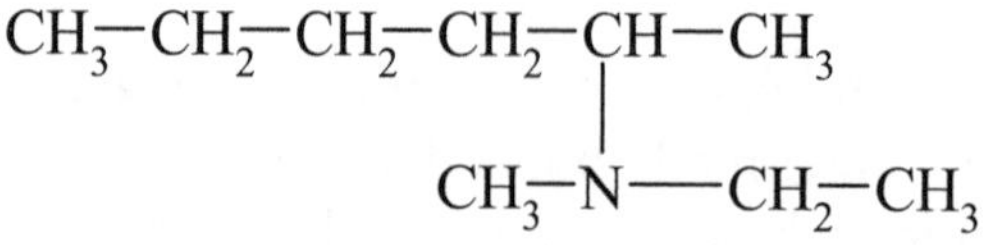 $\underset{\displaystyle CH_3-N-CH_2-CH_3}{CH_3-CH_2-CH_2-CH_2-\overset{|}{C}H-CH_3}$ 3° amine

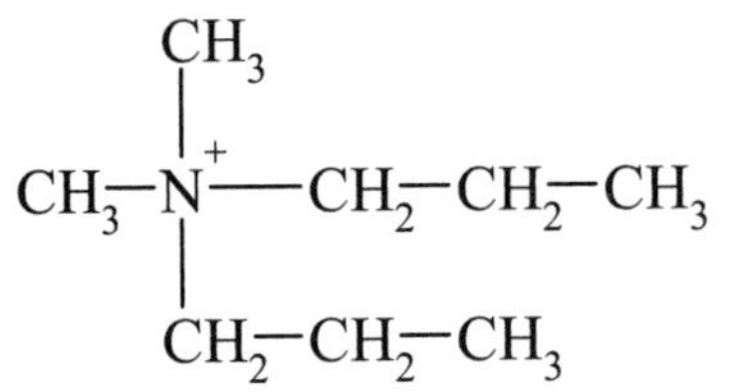

d.   4° amine

8.75   *Identify each compound as a pyridine, a pyrimidine, or a purine (see Figure 8.13a).*

A pyridine has only nitrogen atom replacing a carbon in a benzene ring, a pyrimidine has two, and a purine has two in a benzene ring fused to a five-membered ring containing two additional nitrogen atoms. See text problem for the structures.

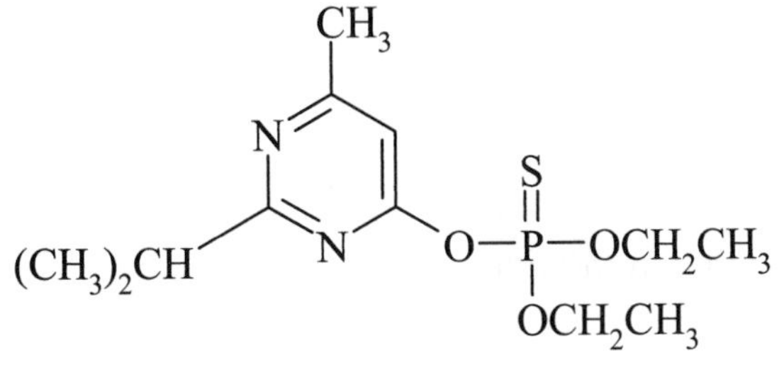

a.   Diazinon (an insecticide)

pyrimidine

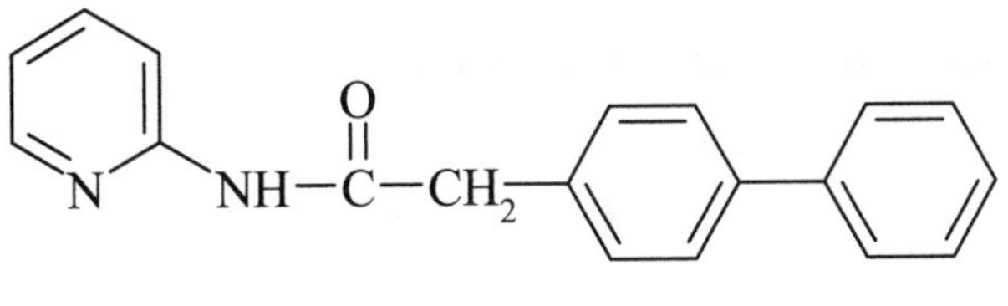

b.   Difenpiramide (an anti-inflammatory drug)

pyridine

8.77   *Account for the fact that ethylamine is more water soluble than hexylamine.*

Although both compounds contain a polar amine group, ethylamine contains a smaller hydrocarbon group than hexylamine, making ethylamine more polar and thus more soluble in water.

8.79   *Draw and name the conjugate acid of*

a. ammonia

$$H-\overset{\overset{\displaystyle H}{|}}{\underset{\underset{\displaystyle H}{|}}{N^+}}-H \qquad \text{ammonium ion}$$

b. propylamine

$$CH_3-CH_2-CH_2-\overset{\overset{\displaystyle H}{|}}{\underset{\underset{\displaystyle H}{|}}{N^+}}-H \qquad \text{propylammonium ion}$$

c. methylethylamine

$$CH_3-\overset{\overset{\displaystyle H}{|}}{\underset{\underset{\displaystyle H}{|}}{N^+}}-CH_2CH_3 \qquad \text{ethylmethylammonium ion}$$

d. triethylamine

$$CH_3-CH_2-\overset{\overset{\displaystyle H}{|}}{\underset{\underset{\displaystyle CH_2CH_3}{|}}{N^+}}-CH_2CH_3 \qquad \text{triethylammonium ion}$$

8.81   *a. Write the equilibrium equation for the acid-base reaction that takes place between propylamine and water.*

Propylamine is basic and so water behaves as the Bronsted-Lowry acid:

$$CH_3CH_2CH_2NH_2 \quad + \quad H_2O \leftrightharpoons CH_3CH_2CH_2NH_3^+ \quad + \quad OH^-$$

b. The conjugate acid of propylamine has a pK$_a$ of 10.6. Which predominates at pH 7, propylamine or its conjugate acid?

Its conjugate acid. The pH of the solution (7) is less than the pKa of the conjugate acid of propylamine (10.6).

8.83 *a. Write a balanced equation for the reaction of propylamine with HCl.*

$$CH_3CH_2CH_2NH_2 \ + \ HCl \ \leftrightharpoons \ CH_3CH_2CH_2NH_3^+ \ Cl^-$$

b. Name the salt that forms.

$$CH_3CH_2CH_2NH_3^+ \ Cl^-$$

propylammonium chloride

8.85 *Anabasine hydrochloride, the salt of anabasine (an amine present in tobacco), is sold as an insecticide. Draw this salt. (Hint: The nitrogen atom in the pyridine ring is not basic.)*

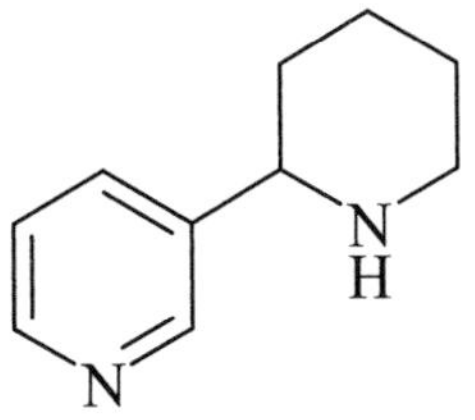

Anabasine

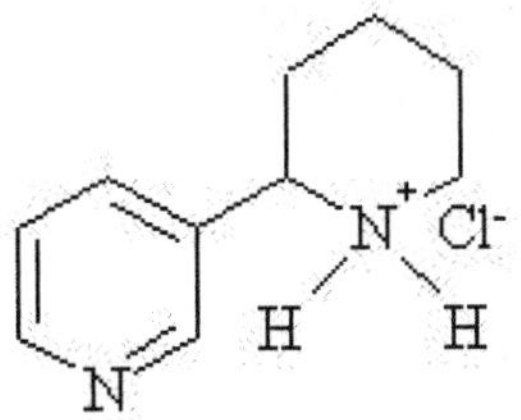

anabasine hydrochloride

8.87 *Caffeine (Figure 8.1c) has four nitrogen atoms, two of which are amines, and the other two of which are amides. Label the amide and amine nitrogen atoms in this compound.*

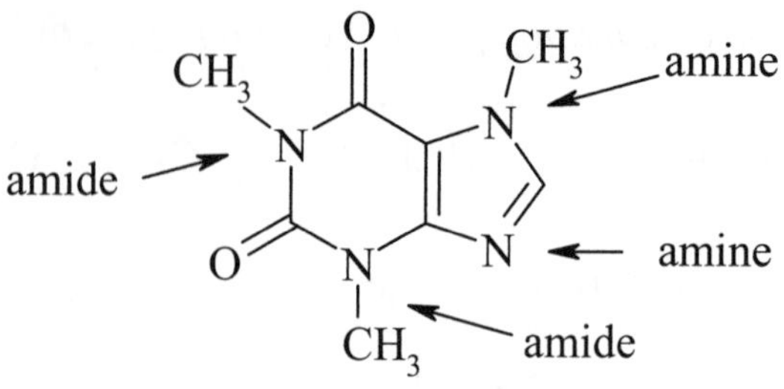

caffeine

8.89   *Draw each amide.*

   a. hexanamide

This is an amide that has 6 carbon atoms in the carboxylic acid residue.

$$CH_3CH_2CH_2CH_2CH_2-\overset{\overset{\displaystyle O}{\|}}{C}-NH_2$$

   b. N-propylacetamide

There is a propyl group bonded to the N atom and there are 2 carbon atoms ("acet") in the carboxylic acid residue.

$$CH_3-\overset{\overset{\displaystyle O}{\|}}{C}-NH-CH_2CH_2CH_3$$

   c. N-butyl-N-methylbenzamide

There is a butyl group and a methyl group bonded to the N atom; the carboxylic acid residue is derived from benzoic acid.

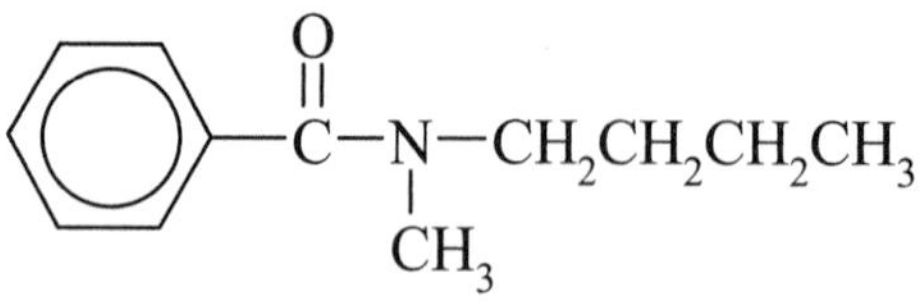

8.91   *Name each amide.*

Use the general IUPAC rules for naming amides. Begin counting from the carbonyl carbon (C=O) to indicate the location of each substituent group on the hydrocarbon chain. Substituent groups bonded to the N atom are identified by placing *N*- before the name of the substituent group. In some cases, a common name is used.

a.

$$\text{C}_6\text{H}_5-\overset{\displaystyle O}{\overset{\|}{C}}-\text{NH}_2$$

benzamide

b.

$$\text{CH}_3-\overset{\displaystyle O}{\overset{\|}{C}}-\underset{\underset{\textstyle \text{CH}_3}{|}}{\text{N}}-\text{CH}_2\text{CH}_3$$

N-ethyl-*N*-methylacetamide

c.

$$\text{CH}_3-\text{CH}_2-\underset{\underset{\textstyle \text{CH}_3}{|}}{\text{CH}}-\overset{\displaystyle O}{\overset{\|}{C}}-\text{NH}_2$$

2-methylbutanamide

8.93   *Draw the amide that will be produced by each reaction.*

When a carboxylic acid reacts with an amine, a bond is formed between the carbonyl group and the nitrogen of the amine group forming an amide.

a.

$$\text{CH}_3-\text{C}_6\text{H}_4-\overset{\displaystyle O}{\overset{\|}{C}}-\text{OH} \quad + \quad \text{NH}_3 \quad \xrightarrow{\text{heat}}$$

The amide product is

$$\boxed{\text{CH}_3-\text{C}_6\text{H}_4-\overset{\displaystyle O}{\overset{\|}{C}}-\text{NH}_2}$$

b.

$$\text{CH}_3-\text{CH}_2-\text{CH}_2-\text{CH}_2-\overset{\displaystyle O}{\overset{\|}{C}}-\text{OH} \quad + \quad \text{NH}_2\text{CH}_3 \quad \xrightarrow{\text{heat}}$$

212

The amide product is

$$CH_3-CH_2-CH_2-CH_2-\overset{\displaystyle O}{\overset{\|}{C}}-NH-CH_3$$

8.95 *Name the amide product of each reaction in Problem 8.93.*

a.

$$CH_3-\underset{}{\bigcirc}-\overset{\displaystyle O}{\overset{\|}{C}}-NH_2 \qquad \text{p-methylbenzamide}$$

b.

$$CH_3-CH_2-CH_2-CH_2-\overset{\displaystyle O}{\overset{\|}{C}}-NH-CH_3 \qquad \textit{N}\text{-methylpentanamide}$$

8.97 *Draw the products formed when each amide in Problem 8.89 is reacted with H₂O in the presence of H⁺.*

The reaction of an amide with water in the presence of H^+ is a hydrolysis reaction. The amide breaks down into a carboxylic acid and a protonated amine.

a.

$$CH_3CH_2CH_2CH_2CH_2\overset{\displaystyle O}{\overset{\|}{C}}-NH_2 \quad + \quad H_2O \quad \xrightarrow{\;H^+\;}$$

$$CH_3CH_2CH_2CH_2CH_2\overset{\displaystyle O}{\overset{\|}{C}}-OH \quad + \quad NH_4^+$$

b.

$$CH_3-\overset{\displaystyle O}{\overset{\|}{C}}-NH\,CH_2CH_2CH_3 \quad + \quad H_2O \quad \xrightarrow{\;H^+\;}$$

$$CH_3-\overset{\displaystyle O}{\overset{\|}{C}}-OH \quad + \quad H-\overset{\displaystyle H}{\underset{\displaystyle H}{\overset{|}{\underset{|}{N^+}}}}-CH_2CH_2CH_3$$

c.

The product reaction at top: benzamide derivative + H_2O with H^+ → benzoic acid + protonated secondary amine (N-butyl-N-methyl).

8.99 *Draw the products formed when each amide in Problem 8.91 is heated in the presence of H_2O and H^+.*

The reaction of an amide with water in the presence of H^+ is a hydrolysis reaction. The amide breaks down into a carboxylic acid and a protonated amine.

a.

b.

c.

8.101 *Draw the products obtained when erbstatin (Problem 8.19) is hydrolyzed under acidic conditions.*

Hydrolysis of erbstatin under acidic conditions results in the breakdown of the amide bond, producing a protonated amine and a carboxylic acid:

8.103 *The boiling point of propanamide is 213°C and that of methyl acetate is 57.5°C. Account for this difference in boiling points for these two molecules with very similar molecular weights.*

Propanamide   Methyl acetate

Boiling points are determined by the strength of intermolecular forces holding the liquid molecules together; the stronger the intermolecular forces, the higher the boiling point for similar size molecules. Propanamide molecules have stronger intermolecular forces because of their ability to hydrogen bond with each other through their –NH groups. Methyl acetate molecules interact only through the weaker dipole-dipole and London intermolecular forces.

8.105 *Olvanil, a compound structurally related to capsaicin (Figure 8.4), has been studied as a potential analgesic. Locate the phenol, amide, and alkene functional groups in the molecule.*

olvanil

215

8.107   *a. What is exfoliation and why is it done?*

Exfoliation is the process of removing the outermost layer of the skin. According to claims made by cosmetic companies, exfoliation "will unblock pores, remove wrinkles and age spots, repair sun damage, and improve the overall feel of the skin."

b. How does mechanical exfoliation differ from chemical exfoliation?

During mechanical exfoliation, the skin is rubbed with abrasive materials to remove the outermost layer. In chemical exfoliation, acids are applied to the skin to peel the outermost layer.

c. How is exfoliation done by a dermatologist similar to exfoliation done at home using skin care products?

Exfoliation done at home and done by dermatologists both make use of acids to chemically remove the outermost layer of the skin.

d. How does exfoliation done by a dermatologist differ from exfoliation done at home using skin care products?

Exfoliation done at home makes use of less concentrated acids than that used by dermatologists.

8.109   *a. Classify each of the compounds in Figure 8.14 as being 1°, 2°, or 3° amine.*

epinephrine (2°)   amphetamine (1°)   methamphetamine (2°)

phenylephrine (2°)   pseudoephedrine (2°)

b. Which functional groups are present in epinephrine, but not in methamphetamine?

-OH group

c. Draw and name the product of the reaction that takes place between ephedrine HCl.

ephedrine hydrochloride

8.111 *According to the Health Link "Adrenaline and Related Compounds," what are some of the biological effects of methamphetamine?*

Methamphetamine acts as a stimulant and has similar, but more potent, effects as amphetamine. These include increased pulse rate and blood pressure.

8.113 *a. What is the function of quorum sensing?*

Quorum sensing is a switch from individual to group activity among bacteria.

b. What role do autoinducers play in quorum sensing?

Autoinducers are signaling molecules. At certain concentrations, autoinducers cause metabolic changes producing compounds that benefit the group initiating quorum sensing.

c. How does biofilm formation benefit bacteria?

The biofilm keeps nutrients close at hand and has channels that expedite water uptake and removal of waste products.

8.115 *Draw the products from the hydrolysis of Lufenuron under acidic conditions.*

8.117   *In addition to capsaicin (Figure 8.4), peppers contain smaller amounts of homocapsaicin and 6,7-dihydrocapsaiciin.*

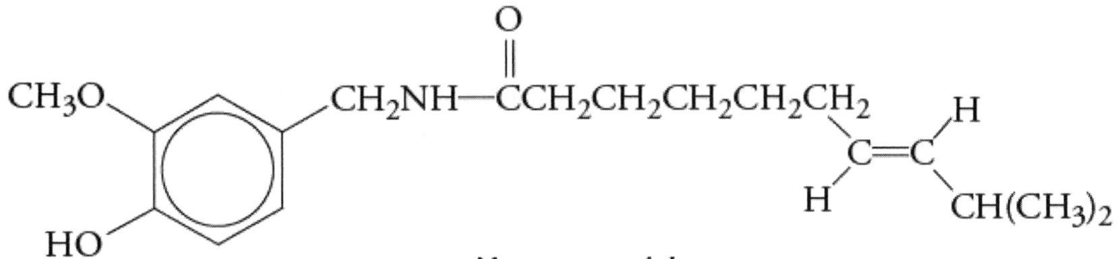

$$CH_3O-\bigcirc-CH_2NH-\overset{\overset{O}{\|}}{C}CH_2CH_2CH_2CH_2CH_2 \quad \underset{H}{\overset{H}{C}}=\underset{CH(CH_3)_2}{C}$$

HO

Homocapsaicin

$$CH_3O-\bigcirc-CH_2NH-\overset{\overset{O}{\|}}{C}CH_2CH_2CH_2CH_2CH_2CH_2CH(CH_3)_2$$

HO

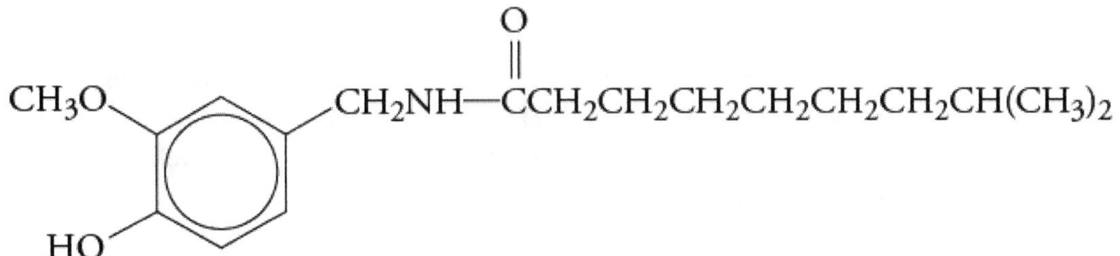

6,7-Dihydrocapsaicin

a. Which best describes the relationship between capsaicin and homocapsaicin? They are: constitutional isomers, different conformations of the same molecule, stereoisomers of one another, identical molecules, entirely different molecules.

Entirely different molecules. The two molecules do not have the same molecular formula. Homocapsaicin has 2 fewer hydrogen atoms.

b. Which stereoisomer is present in homocapsaicin, cis or trans?

Trans

c. Which reaction might you use to convert capsaicin into 6,7-dihydrocapsaicin?

Hydrogenation (across the double bond)

d. Draw the products formed when homcapsaicin is hydrolyzed under acidic conditions.

218

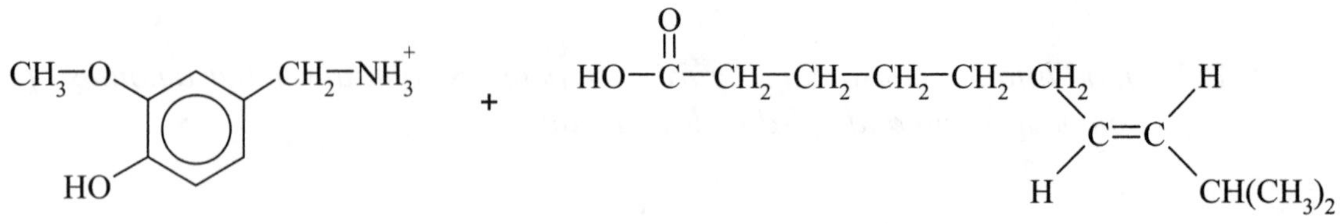

CH_3-O ... $CH_2-NH_3^+$ + $HO-\overset{O}{\overset{\|}{C}}-CH_2-CH_2-CH_2-CH_2-CH_2$ ... $C=C$ ... H, H, $CH(CH_3)_2$

e. Draw the phenoxide salt that forms when 6,7-dihydrocapsaicin is reacted with NaOH.

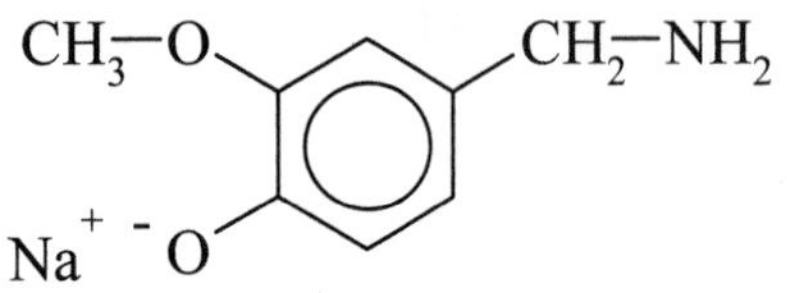

CH_3-O ... CH_2-NH_2

$Na^+ \ ^-O$

8.119 *N-Methylhistamine is one of the products formed during the metabolism of histamine (see the chapter summary). This compound is formed by replacing one of the hydrogen atoms on histamine's primary amine with a methyl group. Draw N- methylhistamine.*

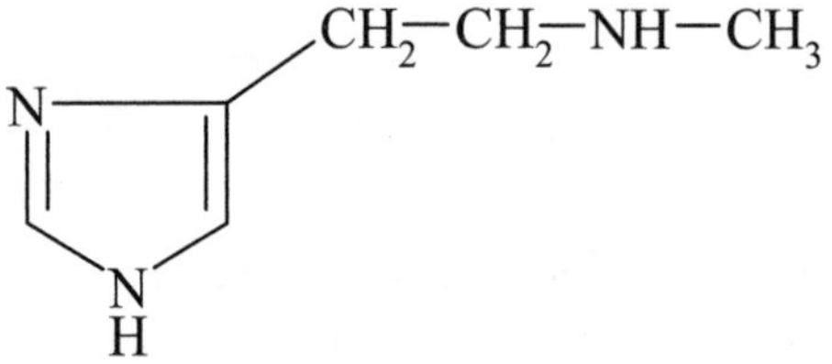

$CH_2-CH_2-NH-CH_3$

N ... N–H

Chapter 9
Alcohols, Ethers, Aldehydes, and Ketones

Solutions to Problems

9.1   *a. Add the missing hydrogen atoms to the ring on each molecule.*

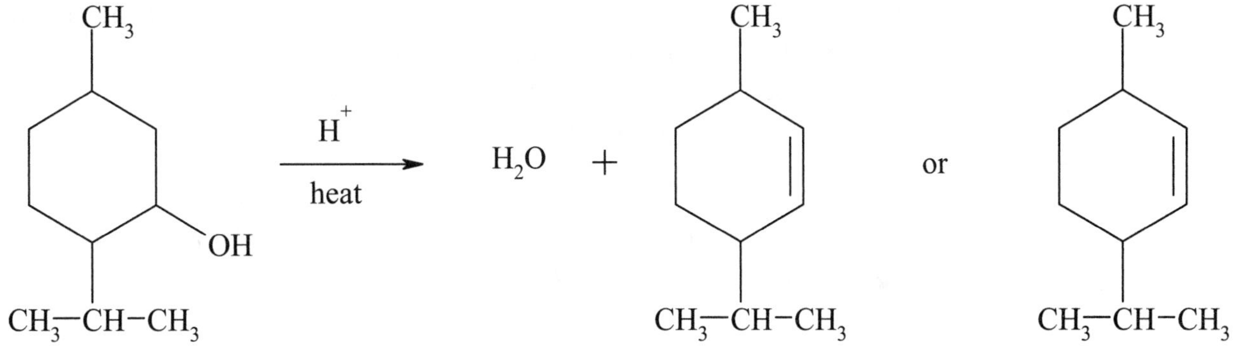

menthol   alkene products

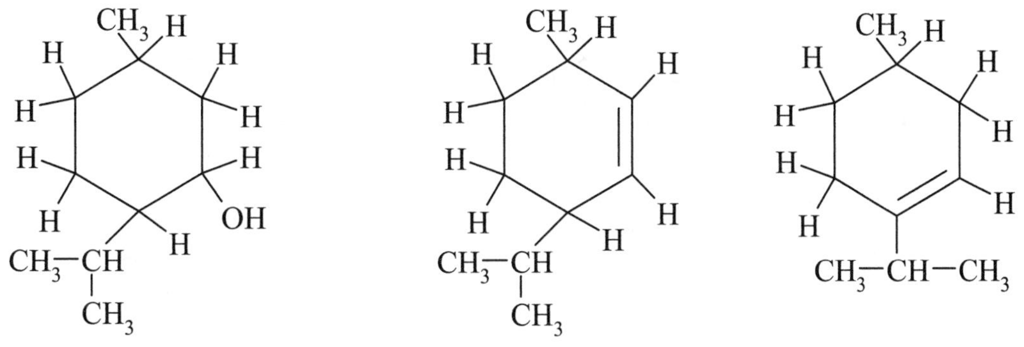

menthol   alkene products

b. When heated in the presence of H^+, menthol reacts to form two alkenes. In the reaction above, which alkene is the major product?

When menthol is heated in the presence of H+, a dehydration reaction occurs. When there are two possible products, the major product is that formed by the removal of –OH from one carbon atom and removal of –H from the neighboring C atom that carries fewer H atoms.

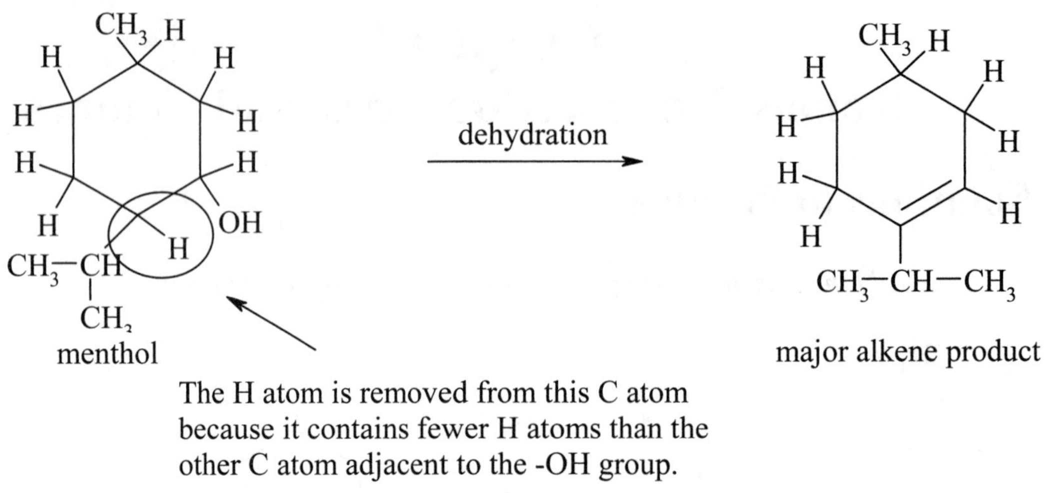

dehydration

menthol

major alkene product

The H atom is removed from this C atom because it contains fewer H atoms than the other C atom adjacent to the -OH group.

c. *Circle the atoms in menthol that are the source of the water molecule that appears as one of the reaction products.*

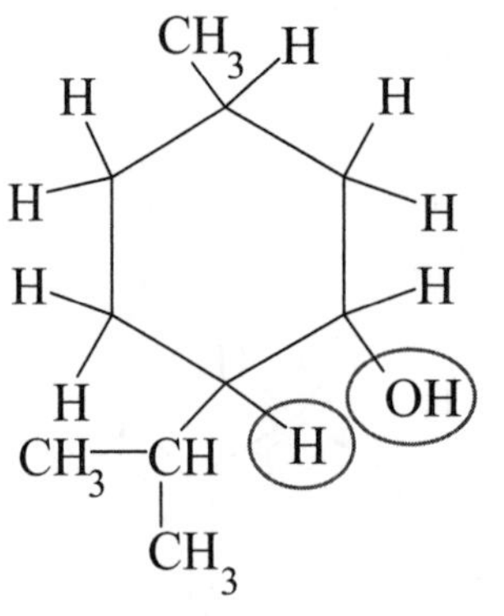

menthol

9.3   *Identify each alcohol as being 1°, 2°, or 3°.*

Alcohols are classified according to the number of carbon atoms bonded to the carbon atom to which the hydroxyl group is attached (C-OH). If there is only C atom bonded to the C-OH, the alcohol is 1°. If there are two C atoms, each one directly bonded to the C-OH, the alcohol is 2°. If there are three C atoms, each one directly bonded to the C-OH, the alcohol is 3°.

a.

$$\underset{\overset{\displaystyle OH}{\displaystyle |}}{CH_3CH_2CH\,CH_2CH_3}$$

2°

b.

$$\underset{\overset{\displaystyle |}{\displaystyle CH_2CH_3}}{\overset{\overset{\displaystyle OH}{\displaystyle |}}{CH_3C\,CH_2CH_3}}$$

3°

c.

$$\underset{}{\overset{\overset{\displaystyle CH_3}{\displaystyle |}}{HOCH_2CH\,CH_2CH_3}}$$

1°

9.5   *Give the IUPAC name of each alcohol in Problem 9.3.*

a. The parent chain has five carbon atoms which makes it a pentane. To name the parent chain drop the "e" and add "ol". The hydroxyl group is on carbon 3 so place a 3- in front of pentanol.

3-pentanol

b. The parent chain has five carbon atoms which makes it a pentane. To name the parent chain drop the "e" and add "ol". The hydroxyl group is on carbon 3 so place a 3- in front of pentanol. There is also a methyl substituent on carbon 3, so assign it the number 3 as well.

3-methyl-3-pentanol

c. The parent chain has four carbon atoms which makes it a butane. To name the parent chain drop the "e" and add "ol". The hydroxyl group is on carbon 1 so place a 1- in front of butanol. There is also a methyl substituent on carbon 2 so assign it the number 2.

2-methyl-1-butanol

9.7   *Draw each alcohol molecule.*

a. 2-hexanol

Draw a six carbon (hexane) parent chain. Add –OH to carbon 2.

$$\underset{CH_3CH_2CH_2CH_2CHCH_3}{\overset{\overset{\displaystyle OH}{|}}{}}$$

b. 3-methyl-1-pentanol

Draw a five carbon (pentane) parent chain. Add –OH to carbon 1. Add –CH$_3$ to carbon 3.

$$\underset{CH_3CH_2CHCH_2CH_2OH}{\overset{\overset{\displaystyle CH_3}{|}}{}}$$

c. 4-isopropylcyclohexanol

Draw a six-membered ring (cyclohexane). Add an –OH. The –OH is carbon 1. Count to carbon 4 and add an isopropyl group.

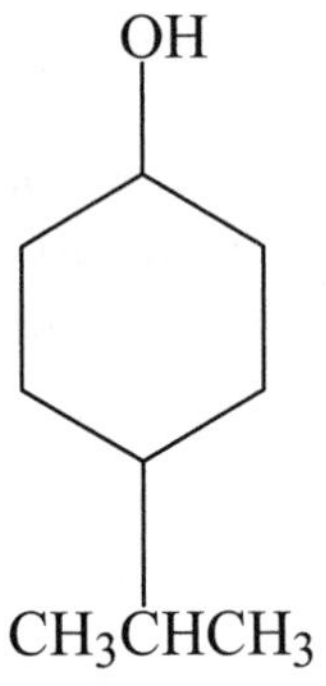

OH

CH$_3$CHCH$_3$

9.9   *Identify each alcohol in Problem 9.7 as being 1°, 2°, or 3°.*

$$\underset{CH_3CH_2CH_2CH_2CHCH_3}{\overset{\overset{\displaystyle OH}{|}}{}}$$

a.   2°

b.

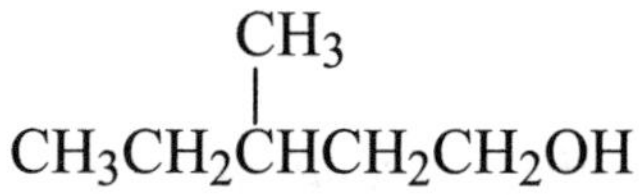

$$CH_3CH_2\overset{\overset{\displaystyle CH_3}{|}}{C}HCH_2CH_2OH$$

1°

c.

OH on top of cyclohexane ring, $CH_3\overset{}{C}HCH_3$ at bottom

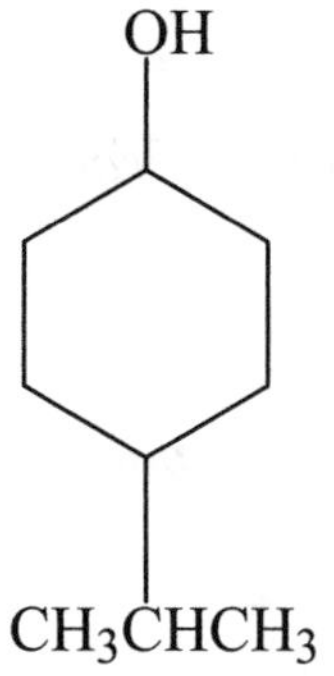

2°

9.11 *To which organic family does each molecule belong?*

   a. $CH_3CH_2SCH_2CH_2CH_3$ sulfides

   b. $CH_3CH_2OCH_2\ CH_2CH_3$ ethers

   c. 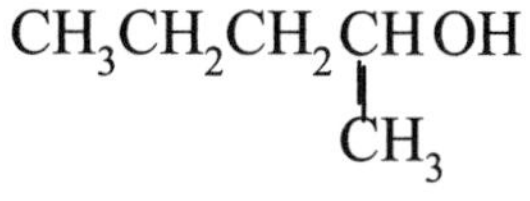 $CH_3CH_2CH_2\overset{\overset{\displaystyle CH_3}{|}}{C}HOH$ alcohols

9.13 *Name each molecule in Problem 9.11*

   a. $CH_3CH_2SCH_2CH_2CH_3$ ethyl propyl sulfide

   b. $CH_3CH_2OCH_2\ CH_2CH_3$ ethyl propyl ether

   c. 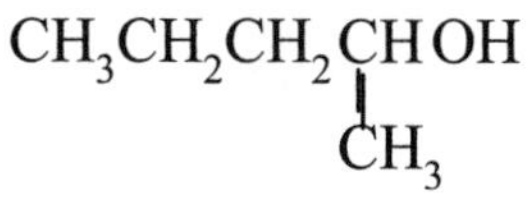 $CH_3CH_2CH_2\overset{\overset{\displaystyle CH_3}{|}}{C}HOH$ 2-pentanol

9.15 *Are the compounds in parts a and b of Problem 9.11 identical molecules, different conformations of the same molecule, cis/trans isomers, constitutional isomers, or entirely different molecules?*

   Entirely different molecules. $CH_3CH_2SCH_2CH_2CH_3$ is a sulfide and $CH_3CH_2OCH_2CH_2CH_3$ is an ether.

9.17   a. What is the shape around the S atom for the molecule in Problem 9.11a?

bent

b. What is the shape around the O atom for the molecule in Problem 9.11b?

 bent

c. One of the molecules in parts a and b of Problem 9.11 is polar and the other is not. Which is which?

a. is not polar, b. is polar

9.19   Give the common name of each ether.

a. 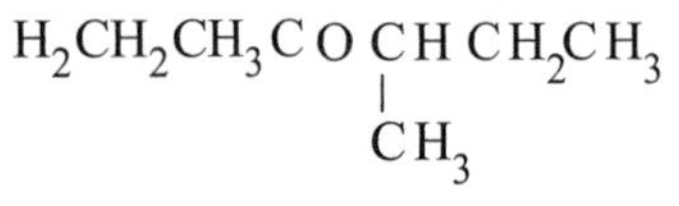 $H_2CH_2CH_3COCHCH_2CH_3$
                          $|$
                          CH_3

   s-butyl propyl ether

b. 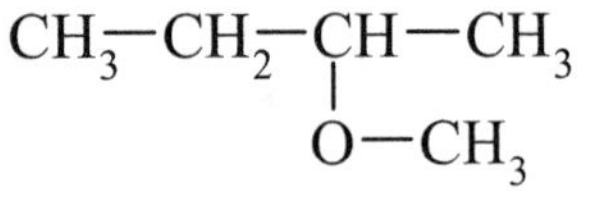 $CH_3-CH_2-CH-CH_3$
                   $|$
                   $O-CH_3$

s-butyl methyl ether

c. 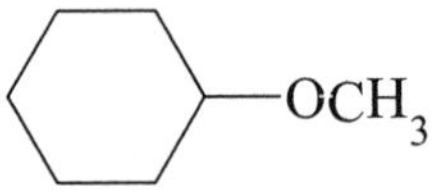   —OCH$_3$

cyclohexyl methyl ether

9.21   Draw each molecule.

a.dimethyl ether

CH_3-O-CH_3

b. dicyclopropyl ether

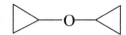

c. butyl ethyl ether

$$CH_3CH_2CH_2CH_2{-}O{-}CH_2CH_3$$

9.23 *The molecule $CH_3CH_2CH_2SH$ is partly responsible for the strong odor that is associated with onions. Give its IUPAC and common name.*

IUPAC name:	1-propanethiol
Common name:	propyl mercaptan

9.25 *a. Allyl mercaptan is the major compound that can be detected on the breath immediately after eating garlic. Draw this compound (allyl = $-CH_2CH{=}CH_2$).*

$$H_2C{=}CH{-}CH_2{-}SH$$

b. Give the IUPAC name for allyl mercaptan.

2-propen-1-thiol

9.27 *Which of the following is a liquid at STP: CH_3CH_2OH or CH_3OCH_3?(see table 9.1)*

CH_3CH_2OH is a liquid at room temperature. CH_3CH_2OH has an –OH group that can hydrogen bond with other molecules of CH_3CH_2OH, leading to strong intermolecular attractive forces. CH_3OCH_3 does not have hydrogen bonding capability, has a low boiling point, and is a gas at STP (see table 9.1).

9.29 *Explain why $CH_3CH_2OCH_2CH_3$ is less soluble in water than its constitutional isomer $CH_3CH_2CH_2CH_2OH$.*

$CH_3CH_2CH_2CH_2OH$ has a greater capability to hydrogen bond with water. This enhances its solubility in the solvent.

9.31   *Ethanol (molecular weight = 46.0 amu) has a boiling point of 75.5°C and propane (molecular weight = 44.0 amu) has a boiling point of -42°C. Account for the difference in boiling point.*

Ethanol is composed of polar molecules that are capable of hydrogen-bonding to each other. Propane is composed of nonpolar molecules held together only by weaker London forces. The stronger intermolecular forces between ethanol molecules cause its boiling point to be much higher than propane's.

9.33   *1-Propanol has a boiling point of 97.4°C and 1-butanol has a boiling point of 117.3°C. Account for the difference in boiling point.*

Both have polar molecules capable of hydrogen-bonding to one another. But butanol has stronger intermolecular London forces due to its longer hydrocarbon chain.

9.35   *Methanethiol (molecular weight = 48.0 amu) has a boiling point of 6°C and ethanol (molecular weight = 46.0 amu) has a boiling point of 78.5°C. Account for the difference in boiling point.*

Although both ethanol and methanethiol contain polar molecules, ethanol molecules are capable of forming the stronger hydrogen-bonding intermolecular force and thus ethanol has the higher boiling point.

9.37   *Isoimpinellin, a naturally occurring ether, is found at very low levels in celery. Tests have shown that at high concentrations this compound can be a carcinogen (cancer-causing agent). This is not cause to avoid eating celery, however, because many foods naturally contain toxic substances at very low levels.*

Isoimpinellin

a. How many ether groups does isoimpinellin contain?

Isoimpinellin has 3 ether groups:

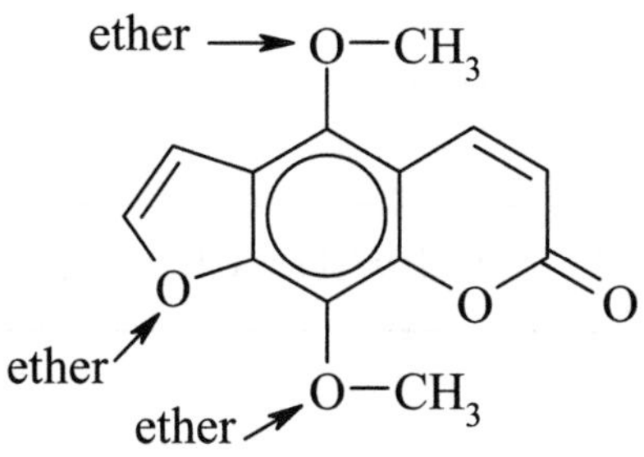

ether → O—CH₃

ether

ether → O—CH₃

b. What other functional groups are present in this molecule?

alkene, aromatic, ester

9.39 *Draw the products of each nucleophilic substitution reaction.*

a. 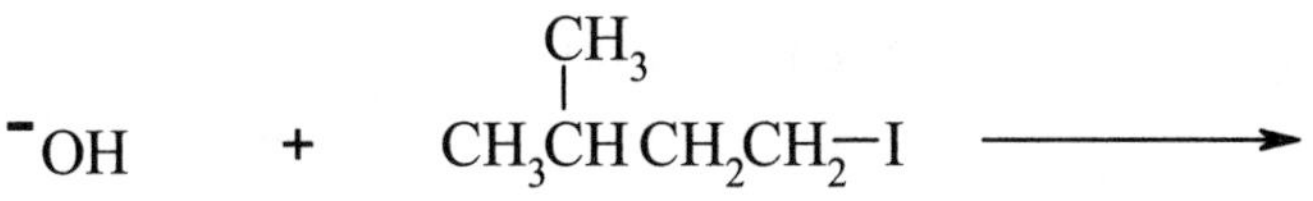 ⁻OH   +   CH₃CH CH₂CH₂—I   ⟶

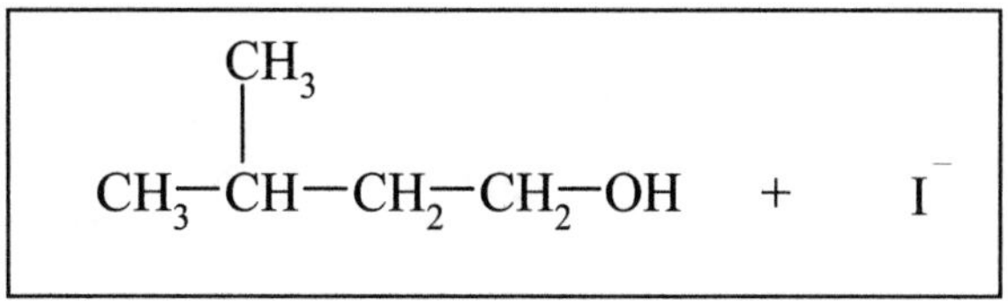

| CH₃ |
| CH₃—CH—CH₂—CH₂—OH   +   I⁻ |

b. ⁻OCH₂CH₂CH₃   +   CH₃—Br   ⟶

| CH₃—O—CH₂—CH₂—CH₃   +   Br⁻ |

c. 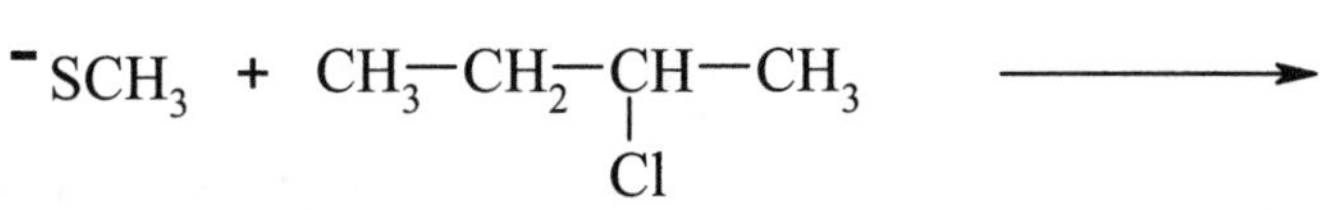 ⁻SCH₃ + CH₃—CH₂—CH—CH₃   ⟶
　　　　　　　　　　　　　|
　　　　　　　　　　　　 Cl

| CH₃—CH₂—CH—CH₃   +   Cl⁻ |
| 　　　　　　| |
| 　　　　　S—CH₃ |

228

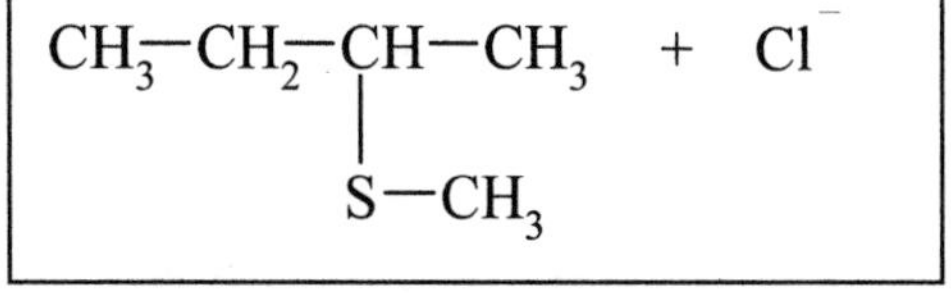

d. $^-$SH + $CH_3-CH_2-CH_2-CH_2-Br$ $\longrightarrow$

$$\boxed{CH_3-CH_2-CH_2-CH_2-SH \quad + \quad Br^-}$$

9.41 *Name each of the products in Problem 9.39.*

    a. 3-methyl-1-butanol
    b. methyl propyl ether
    c. s-butyl methyl sulfide
    d. 1-butanethiol or butyl mercaptan

9.43 *Draw the missing alkyl bromide for each reaction.*

  a. $^-OCH_3$ + $\boxed{CH_3-CH_2-Br}$ $\longrightarrow$ $CH_3-CH_2-O-CH_3$

  b. ^-SH + $\boxed{CH_3-CH_2-CH_2-Br}$ $\longrightarrow$ $CH_3-CH_2-CH_2-SH$

  c. $^-SCH_2CH_3$ + $\boxed{CH_3-CH_2-Br}$ $\longrightarrow$ $CH_3-CH_2-S-CH_2-CH_3$

9.45 *Write a chemical reaction that shows how each compound can be produced from an alkyl bromide.*

  a. CH₃CH₂CH₂OH

  ^-OH + $CH_3-CH_2-CH_2-Br$ $\longrightarrow$ $CH_3-CH_2-CH_2-OH$

b. $CH_3OCH_2CH_2CH_3$

$^-OCH_2CH_2CH_3$ + CH_3-Br $\longrightarrow$ $CH_3O-CH_2-CH_2-CH_3$

or

$^-OCH_3$ + $CH_3CH_2CH_2-Br$ $\longrightarrow$ $CH_3O-CH_2-CH_2-CH_3$

9.47 *Draw the major product of each reaction.*

These reactions are hydration reactions in which water adds to the double bond. According to Markovnikov's rule, the major product is the one in which the hydrogen atom from the water molecule makes a bond with the carbon that has the higher number of hydrogen atoms and the –OH group makes a bond with the carbon that has fewer hydrogen atoms.

a. $CH_3-CH_2-\underset{\underset{CH_3}{|}}{C}=CH_2$ + H_2O $\xrightarrow{H^+}$

$$CH_3-CH_2-\underset{\underset{CH_3}{|}}{\overset{\overset{OH}{|}}{C}}-CH_3$$

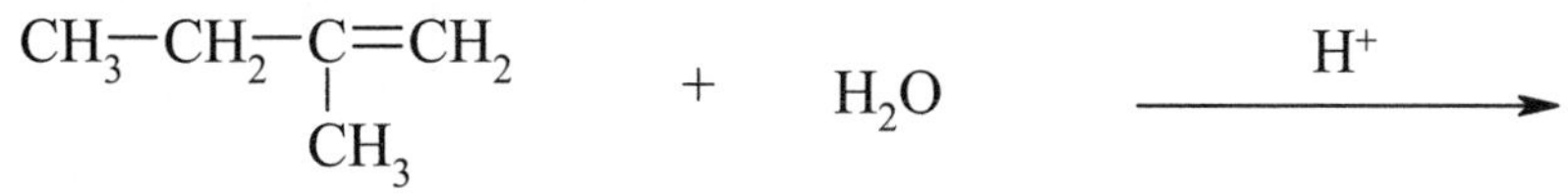

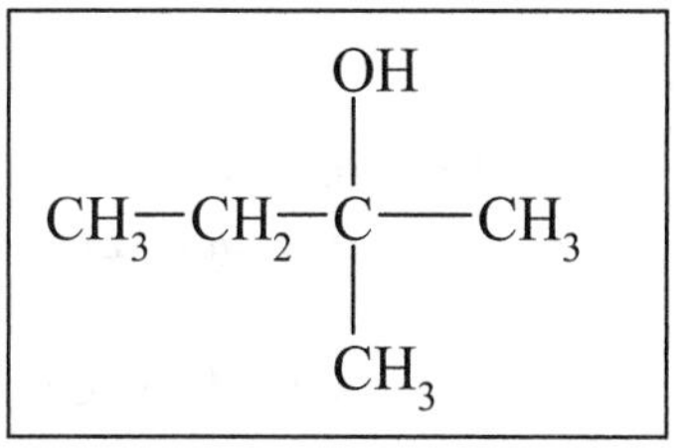

b.

+ H_2O $\xrightarrow{H^+}$

230

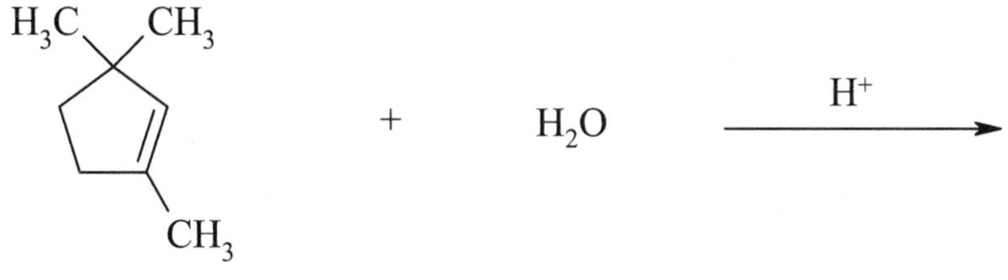

c. $CH_2{=}C{-}CH_2{-}CH_2{-}CH_3$ + H_2O $\xrightarrow{\ H^+\ }$

        |
       CH_3

9.49   *a. When a particular alkene is reacted with H_2O and H^+, 2-butanol is the only product. Draw and name this alkene.*

    $CH_3{-}CH{=}CH{-}CH_3$ 2-butene

   b. When a different alkene is reacted with H_2O and H^+, 2-butanol is the major product. Draw and name this alkene.

    $CH_2{=}CH{-}CH_2{-}CH_3$ 1-butene

9.51   *The molecule CH_3Cl is named methyl chloride. Name each of the following molecules:*

These are all non-IUPAC names:

   a. CH_3CH_2Br   ethyl bromide

   b. CH_3CHFCH_3   isopropyl fluoride

   c. $CH_3CH_2CH_2CH_2I$ butyl iodide

d. CH₃CH(CH₃)CH₂Cl   isobutyl chloride

9.53   *Draw the organic product (if any) expected from each reaction.*

Alcohols undergo oxidation to aldehydes or carboxylic acids or ketones when they react with $K_2Cr_2O_7$. When thiols are oxidized by I_2, two molecules of the thiol react to form a disulfide bond.

a. 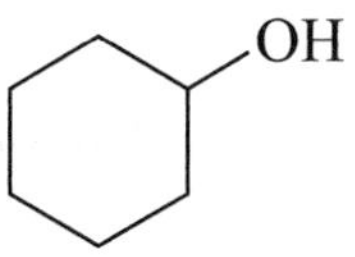   [cyclohexanol structure with —OH]   + $K_2Cr_2O_7$ ⟶

[boxed: cyclohexanone structure with =O]

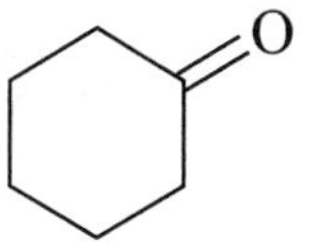

b. $HO-CH_2-CH_2-CH_2-CH_2-CH_3$ + $K_2Cr_2O_7$ ⟶

[boxed:
$$HO-\overset{\overset{\text{O}}{\|}}{C}-CH_2-CH_2-CH_2-CH_3$$
]

c. 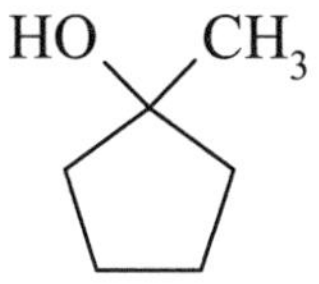   [cyclopentane structure with HO and CH₃]   + $K_2Cr_2O_7$ ⟶

[boxed: no reaction (3° alcohol)]

d.   2 CH_3-SH + I_2 ⟶

[boxed: $CH_3-S—S-CH_3$]

e. 2 CH$_3$-CH-CH$_3$ +   I$_2$ ⟶
         |
         SH

$$\boxed{\begin{array}{c} \text{CH}_3\text{-CH-CH}_3 \\ | \\ \text{S} \\ | \\ \text{S} \\ | \\ \text{CH}_3\text{-CH-CH}_3 \end{array}}$$

9.55   *Describe the difference in the products obtained when 1-butanol and 1-butanethiol are oxidized.*

When 1-butanol is oxidized, a carboxylic acid is produced. When 1-butanethiol is oxidized, a disulfide is produced.

9.57   *a. Lactate, which builds up in muscle cells during exercise, is sent to the liver where an enzyme catalyzes the oxidation of this 2° alcohol to pyruvate. Draw pyruvate.*

CH$_3$-CH-C-O$^-$ +   NAD$^+$ $\xrightarrow{\text{enzyme}}$ +   NADH + H$^+$
    | ||
    HO   O

Lactate   Pyruvate

When oxidation takes place, the 2° alcohol group is converted to a ketone:

CH$_3$-C-C-O$^-$
    || ||
    O   O

pyruvate

b. Lactate and pyruvate are the conjugate bases of lactic acid and pyruvic acid, respectively. Draw these carboxylic acids.

CH$_3$-CH-C-OH   CH$_3$-C-C-OH
    | || || ||
    OH O   O   O

lactic acid   pyruvic acid

c. At pH 7, why does each of these acids appear in its conjugate base form?

pH = 7 is greater than the pKa of a typical carboxylic acid.

233

9.59   *Draw the major product of each reaction.*

In the presence of heat and H^+, the –OH group of an alcohol is removed, along with a neighboring hydrogen, to form a double bond. The hydrogen atom removed is from the neighboring carbon atom that carries the fewest H atoms.

a. $CH_3CH\,CH\,CH\,CH_3$ (with CH_3 above second carbon, CH_3 and OH below) $\xrightarrow[\text{heat}]{H+}$

$$\boxed{CH_3CHC{=}CHCH_3 \text{ (with } CH_3 \text{ above and } CH_3 \text{ below)}}$$

b. $CH_3C\,CH_2CH_3$ (with CH_3 above, OH below) $\xrightarrow[\text{heat}]{H+}$

$$\boxed{CH_3C{=}CHCH_3 \text{ (with } CH_3 \text{ above)}}$$

9.61   *Draw each molecule named below and draw the major organic product expected when each is reacted with H^+ and heat.*

In the presence of heat and H^+, the –OH group of an alcohol is removed, along with a neighboring hydrogen, to form an alkene. The hydrogen atom removed is from the neighboring carbon atom that carries the fewest H atoms.

a. 2,3-dimethyl-2-butanol

$CH_3CH{-}CCH_3$ (with CH_3 and OH above, CH_3 below)   yields $CH_3C{=}CCH_3$ (with CH_3 above, CH_3 below)

b. 2,3-dimethyl-3-hexanol

$$\underset{\underset{\displaystyle CH_3}{|}}{\overset{\overset{\displaystyle OH \quad CH_3}{| \quad\quad |}}{CH_3CH_2CH_2C - CHCH_3}} \qquad \text{yields} \qquad \underset{\underset{\displaystyle CH_3}{|}}{\overset{\overset{\displaystyle CH_3}{|}}{CH_3CH_2CH_2C = CHCH_3}}$$

c. *2-methylcyclopentanol*

OH

—CH3

yields

—CH3

9.63   *a. When a particular alcohol is heated in the presence of H^+, 2-methyl-2 butene is the major product. Draw and name this alcohol.*

In the presence of heat and H^+, the –OH group of an alcohol is removed, along with a neighboring hydrogen, to form an alkene. The hydrogen atom removed is from the neighboring carbon atom that carries the fewest H atoms.

$$\underset{\underset{\displaystyle CH_3}{|}}{\overset{\overset{\displaystyle OH}{|}}{CH_3 - C - CH_2 - CH_3}} \qquad \xrightarrow{\quad H^+,\ heat \quad} \qquad \underset{\underset{\displaystyle CH_3}{|}}{CH_3 - C = CH - CH_3}$$

2-methyl-2-butanol

2-methyl-2-butene (major product)

b. When a different alcohol is heated in the presence of H^+, 2-methyl-2 butene is the major product. Draw and name this alcohol.

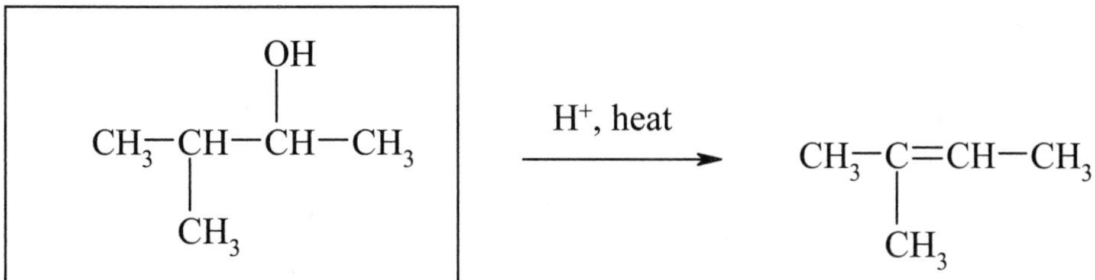

OH
|
$CH_3-CH-CH-CH_3$
|
CH_3

3-methyl-2-butanol

$\xrightarrow{\text{H}^+,\ \text{heat}}$

$CH_3-C=CH-CH_3$
|
CH_3

2-methyl-2-butene (major product)

9.65   *Name each of the following molecules.*

CH_3 O
        | ||
a. 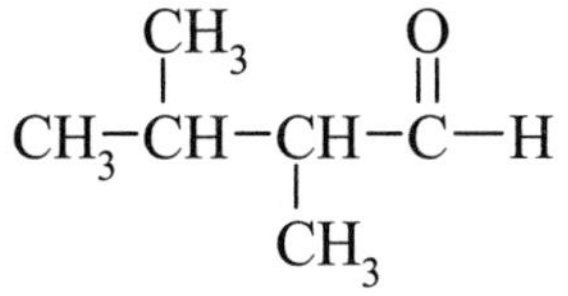 $CH_3-CH-CH-C-H$
                   |
                 CH_3

2,3-dimethylbutanal

         O
         ||
b. 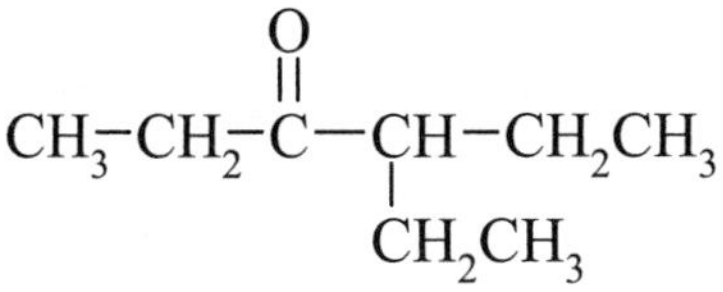 $CH_3-CH_2-C-CH-CH_2CH_3$
                      |
                    CH_2CH_3

4-ethyl-3-hexanone

   CH_3-CH_2 O
              | ||
c. 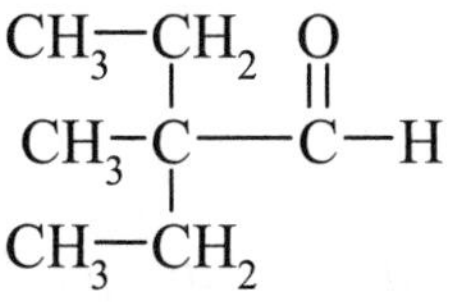 $CH_3-C-----C-H$
              |
   CH_3-CH_2

2-ethyl-2-methylbutanal

      O CH_3
      || |
d. 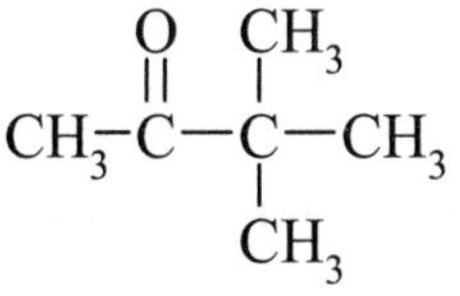 $CH_3-C-C-CH_3$
              |
            CH_3

3,3-dimethyl-2-butanone

236

9.67 *Draw each molecule.*

a. pentanal

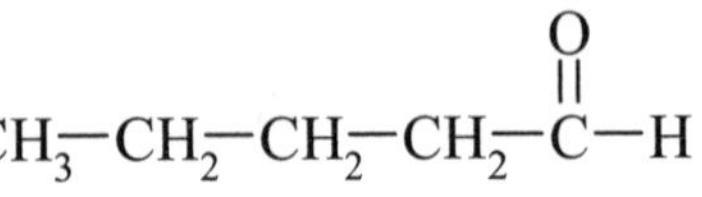

$$CH_3-CH_2-CH_2-CH_2-\overset{\overset{\displaystyle O}{\|}}{C}-H$$

b. 3-bromohexanal

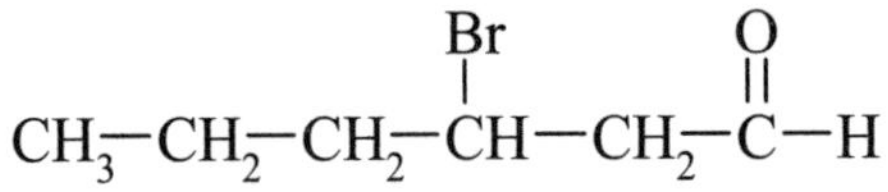

$$CH_3-CH_2-CH_2-\overset{\overset{\displaystyle Br}{|}}{CH}-CH_2-\overset{\overset{\displaystyle O}{\|}}{C}-H$$

c. dipropyl ketone

$$CH_3-CH_2-CH_2-\overset{\overset{\displaystyle O}{\|}}{C}-CH_2-CH_2-CH_3$$

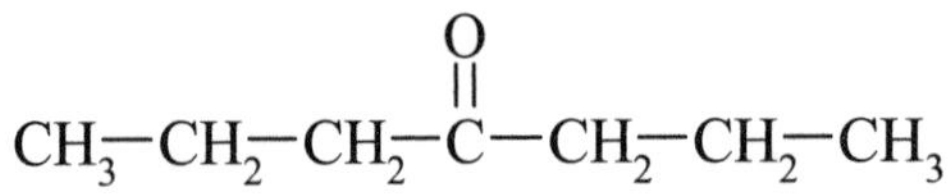

d. 2,5-dibromocyclohexanone

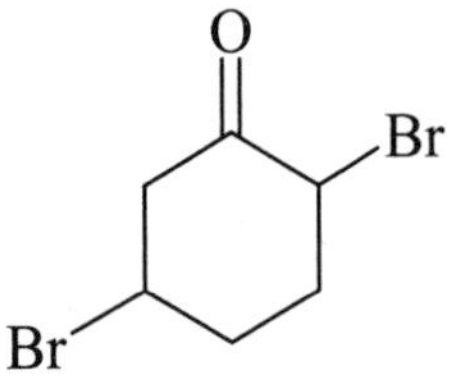

9.69 *2-Hexanone has a boiling point of 150°C and 2-pentanone has a boiling point of 102°C.*

a. Draw each compound.

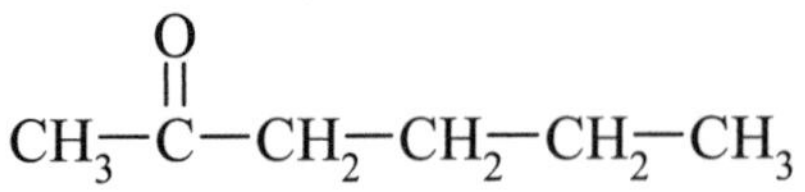

$$CH_3-\overset{\overset{\displaystyle O}{\|}}{C}-CH_2-CH_2-CH_2-CH_3 \qquad CH_3-\overset{\overset{\displaystyle O}{\|}}{C}-CH_2-CH_2-CH_3$$

2-hexanone 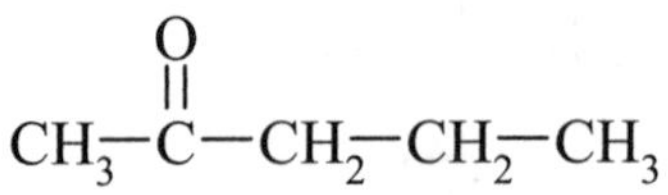   2-pentanone

b. Account for the difference in boiling point.

2-hexanone has a higher boiling point than 2-pentanone because it is a larger molecule and so stronger London dispersion forces exist between its molecules.

9.71   *Propanone (molecular weight = 58.0 amu) has a boiling point of 56°C and 1-propanol (molecular weight = 60.0 amu) has a boiling point of 97.4°C. Account for the difference in boiling point.*

1-propanol molecules are able to hydrogen-bond whereas propanone molecules are not.

9.73   *Draw the product (if any) of each reaction.*

a.

$$CH_3\!-\!CH_2\!-\!\underset{\underset{CH_3}{|}}{CH}\!-\!CH_2\!-\!\overset{\overset{O}{\|}}{C}\!-\!H \xrightarrow{K_2Cr_2O_7} \boxed{CH_3\!-\!CH_2\!-\!\underset{\underset{CH_3}{|}}{CH}\!-\!CH_2\!-\!\overset{\overset{O}{\|}}{C}\!-\!OH}$$

b.

$$HO\!-\!CH_2\!-\!CH_2\!-\!CH_2\!-\!\underset{\underset{CH_3}{|}}{\overset{\overset{CH_3}{|}}{C}}\!-\!CH_3 \xrightarrow{K_2Cr_2O_7} \boxed{HO\!-\!\overset{\overset{O}{\|}}{C}\!-\!CH_2\!-\!CH_2\!-\!\underset{\underset{CH_3}{|}}{\overset{\overset{CH_3}{|}}{C}}\!-\!CH_3}$$

c.

$$CH_3\!-\!\underset{\underset{H}{|}}{\overset{\overset{OH}{|}}{C}}\!-\!\underset{\underset{CH_3}{|}}{CH}\!-\!\overset{\overset{O}{\|}}{C}\!-\!H \xrightarrow{K_2Cr_2O_7} \boxed{CH_3\!-\!\overset{\overset{O}{\|}}{C}\!-\!\underset{\underset{CH_3}{|}}{CH}\!-\!\overset{\overset{O}{\|}}{C}\!-\!OH}$$

9.75   *Draw each molecule named below. Draw the product (if any) obtained when each is reacted with $K_2Cr_2O_7$.*

a. propanal

When an aldehyde is oxidized by $K_2Cr_2O_7$ the product is a carboxylic acid.

$$CH_3CH_2\overset{\overset{O}{\|}}{C}\!-\!H \quad \text{yields} \quad CH_3CH_2\overset{\overset{O}{\|}}{C}\!-\!OH$$

b. butanone

Ketones are not oxidized by $K_2Cr_2O_7$.

$$\underset{\displaystyle CH_3CH_2\overset{\displaystyle O}{\overset{\displaystyle \|}{C}}CH_3}{}$$ yields no reaction

c. *1-butanol*

First, the alcohol is oxidized by $K_2Cr_2O_7$ to an aldehyde. Then the aldehyde is oxidized to a carboxylic acid.

$$CH_3CH_2CH_2CH_2OH \quad \text{yields} \quad CH_3CH_2CH_2\overset{\displaystyle O}{\overset{\displaystyle \|}{C}}-OH$$

d. *3-chloro-3-methylpentanal*

When an aldehyde is oxidized by $K_2Cr_2O_7$ the product is a carboxylic acid.

$$\underset{\displaystyle \underset{\displaystyle Cl}{|}}{CH_3CH_2\overset{\displaystyle \overset{\displaystyle CH_3}{|}}{C}HCH_2\overset{\displaystyle O}{\overset{\displaystyle \|}{C}}-H} \quad \text{yields} \quad \underset{\displaystyle \underset{\displaystyle Cl}{|}}{CH_3CH_2\overset{\displaystyle \overset{\displaystyle CH_3}{|}}{C}HCH_2\overset{\displaystyle O}{\overset{\displaystyle \|}{C}}-OH}$$

e. *3-hydroxybutanal*

When an aldehyde is oxidized by $K_2Cr_2O_7$ the product is a carboxylic acid. This molecule also has a secondary alcohol group which is oxidized to a ketone.

$$CH_3\overset{\displaystyle \overset{\displaystyle OH}{|}}{C}HCH_2\overset{\displaystyle O}{\overset{\displaystyle \|}{C}}-H \quad \text{yields} \quad CH_3\overset{\displaystyle O}{\overset{\displaystyle \|}{C}}CH_2\overset{\displaystyle O}{\overset{\displaystyle \|}{C}}-OH$$

9.77 *Draw the product of each reaction.*

When aldehydes are reacted with H_2 in the presence of a Pt catalyst, the aldehyde is converted to an alcohol.

a. $\underset{\displaystyle \underset{\displaystyle CH_2CH_3}{|}}{CH_3CH_2\overset{\displaystyle O}{\overset{\displaystyle \|}{C}}CHCH_3}$ + H_2 $\xrightarrow{\text{Pt}}$

239

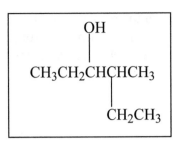

b.   CH₃CH₂⟨cyclopentanone with ethyl⟩=O + H₂ $\xrightarrow{Pt}$

CH₃CH₂⟨cyclopentane ring⟩—OH

c.   CH₃C̈CH₂C̈—H   +   H₂ $\xrightarrow{Pt}$ CH₃CCH₂CH₂OH

with CH₃ and O above, CH₃ below

CH₃ | CH₃CCH₂CH₂OH | CH₃

9.79   *One step in the synthesis of fatty acids by the body involves the enzyme catalyzed reduction of a ketone group present in a carboxylic acid derivative. In this compound, "ACP" is acyl carrier protein, which will be discussed in Chapter 14. Draw the reduction product.*

$$CH_3-CH_2-CH_2-\overset{O}{\overset{\|}{C}}-CH_2-\overset{O}{\overset{\|}{C}}-ACP + NADPH + H^+ \xrightarrow{enzyme} ? + NADP^+$$

Reduction converts the alcohol groups into ketone groups:

OH   OH

CH₃CH₂CH₂CH–CH₂CH- ACP

9.81   *Draw the product of each reaction.*

In the presence of H^+ as a catalyst, alcohols can undergo an addition reaction with the carbonyl group of an aldehydes or ketone to form hemiacetals. When an aldehyde or a ketone reacts with 2 molecules of the alcohol, an acetal is formed.

a.

$$CH_3-CH_2-CH_2-\overset{\overset{\textstyle O}{\|}}{CH} \ + \ CH_3-\underset{\underset{\textstyle CH_3}{|}}{CH}-OH \ \underset{}{\overset{H^+}{\rightleftharpoons}}$$

$$\boxed{CH_3-CH_2-CH_2-\underset{\underset{\textstyle CH_3}{\underset{|}{O-CH-CH_3}}}{\overset{\overset{\textstyle OH}{|}}{C}}-H}$$

b.

$+ \ 2 \quad CH_3-CH_2-CH_2-CH_2-OH \quad \overset{H^+}{\rightleftharpoons}$

$$\boxed{CH_3CH_2CH_2CH_2O \diagdown \diagup OCH_2CH_2CH_2CH_3}$$

c.

$$CH_3-CH_2-\underset{\underset{\textstyle CH_3}{|}}{CH}-\overset{\overset{\textstyle O}{\|}}{C}-CH_2-CH_2-CH_3 \ + \ \text{(cyclopentyl)}-OH \ \overset{H^+}{\rightleftharpoons}$$

$$\boxed{CH_3-CH_2-\underset{\underset{\textstyle CH_3 \ \ OH}{|}}{CH}-\overset{\overset{\textstyle O-\text{(cyclopentyl)}}{|}}{C}-CH_2-CH_2-CH_3}$$

241

9.83 *Draw the missing reactant for each reaction.*

a. 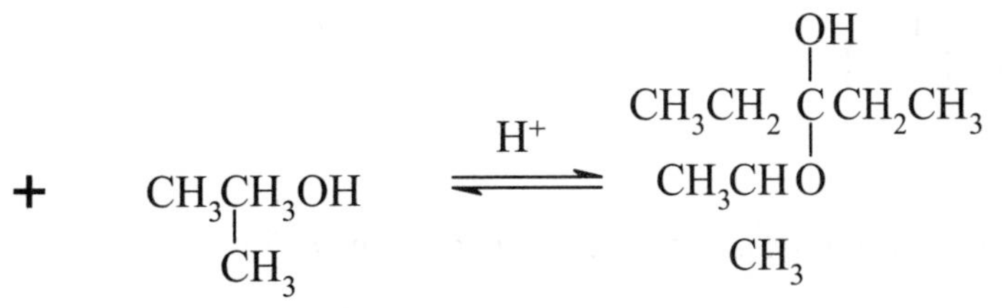
$$ + \quad \underset{\underset{CH_3}{|}}{CH_3CH_2OH} \quad \underset{\longleftarrow}{\overset{H^+}{\rightleftharpoons}} \quad \underset{\underset{CH_3}{|}}{\overset{\overset{OH}{|}}{CH_3CH_2\,\underset{CH_3CHO}{C}CH_2CH_3}} $$

Since the product is a hemiacetal at position 3 of a 5 carbon chain, the original reactant had to be 3-pentanone.

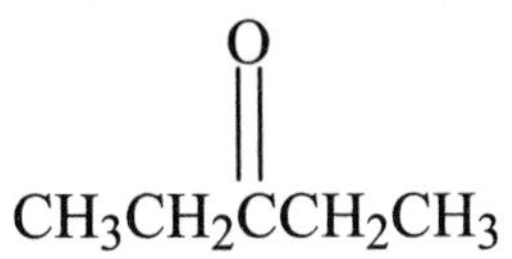

$$ \underset{CH_3CH_2CCH_2CH_3}{\overset{\overset{O}{\|}}{}} $$

b. 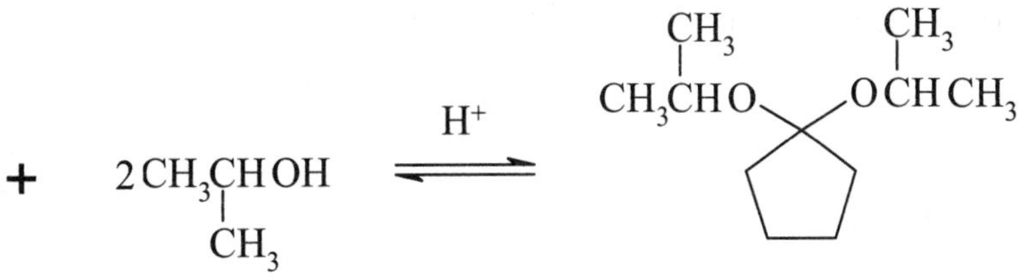
$$ + \quad 2\,\underset{\underset{CH_3}{|}}{CH_3CHOH} \quad \overset{H^+}{\rightleftharpoons} \quad $$

(product: acetal with two $CH_3CHO\text{—}CH_3$ groups on a 5-carbon ring)

Since the product is an acetal formed by addition of two alcohol molecules, the reactant contained C=O in a 5-carbon ring.

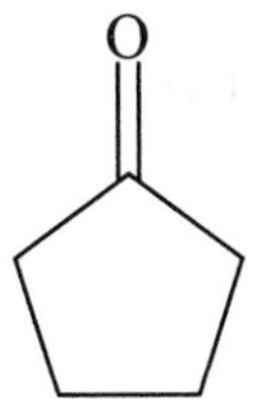

(cyclopentanone structure)

c. 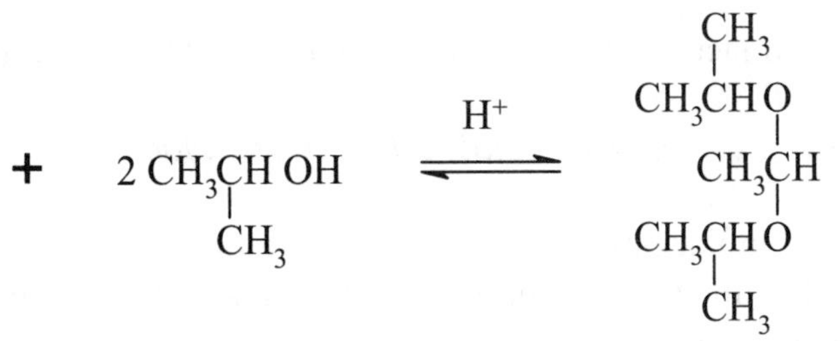
$$ + \quad 2\,\underset{\underset{CH_3}{|}}{CH_3CH\,OH} \quad \overset{H^+}{\rightleftharpoons} \quad \begin{array}{c} \overset{CH_3}{|} \\ CH_3CHO \\ | \\ CH_3CH \\ | \\ CH_3CHO \\ | \\ CH_3 \end{array} $$

Since the product is an acetal formed by addition of two alcohol molecules, the reactant was an aldehyde.

$$\begin{array}{c} O \\ \parallel \\ CH_3C-H \end{array}$$

9.85 *Draw the missing reactant for each reaction.*

a.

$$\begin{array}{c} O \\ \parallel \\ CH_3-C-CH_2-CH_3 \end{array} + \boxed{\begin{array}{c} CH_3-CH-OH \\ | \\ CH_3 \end{array}} \;\overset{H^+}{\rightleftharpoons}\; \begin{array}{c} OH \\ | \\ CH_3-C-CH_2-CH_3 \\ | \\ O-CH-CH_3 \\ | \\ CH_3 \end{array}$$

b.

$$\begin{array}{c} O \\ \parallel \\ CH_3-CH_2-C-CH_2-CH_3 \end{array} + \boxed{CH_3-CH_2-CH_2-OH} \;\overset{H^+}{\rightleftharpoons}\;$$

$$\begin{array}{c} OH \\ | \\ CH_3-CH_2-C-CH_2-CH_3 \\ | \\ O-CH_2-CH_2-CH_3 \end{array}$$

c.

$$\begin{array}{c} O \\ \parallel \\ \bigcirc\!\!-CH \end{array} + \boxed{2\ CH_3-OH} \;\overset{H^+}{\rightleftharpoons}\; \begin{array}{c} O-CH_3 \\ | \\ \bigcirc\!\!-CH \\ | \\ O-CH_3 \end{array}$$

9.87 *a. How are cyclic hemiacetals similar to lactones (8.19)?*

Cyclic hemiacetals and lactones each have a ring that contains a C-O-C linkage.

b. How are cyclic hemiacetals different from lactones?

In cyclic hemiacetals, one of the carbon atoms in the C-O-C linkage is also attached to an –OH group. In lactones, this carbon atom is double-bonded to an oxygen atom (C=O).

9.89 *The toxicity of methanol is mainly due to the aldehyde produced when it is oxidized in the liver. Draw methanol, then draw and name the aldehyde formed on its oxidation.*

methanol   methanal or formaldehyde

$$CH_3OH \qquad\qquad H-\overset{\overset{\displaystyle O}{\displaystyle \|}}{C}-H$$

9.91 *O_2 is reduced when it is converted into H_2O_2. Explain.*

Hydrogen atoms are added to O_2 to form H_2O_2.

9.93 *In the breakdown of superoxide, which product results from the reduction of O_2^- and which results from the oxidation of O_2^-? Explain.*

$$2O_2^- + 2H^+ \rightarrow H_2O_2 + O_2$$

The product that results from reduction of O_2^- is H_2O_2. The change is classified as reduction because it involves a gain of H. The product that results from oxidation of O_2^- is O_2. This change is classified as oxidation because it involves a loss of electrons.

9.97 *The anti-depressant fluoxetine (Prozac), which has been detected in biosolids, is typically sold as the hydrochloride salt. Draw this salt.*

Hydrochloride salt of fluoxetine

9.99 *The anti-depressant Venlefaxine has been detected in ground water at concentrations of 50 µg/L.*

a. *At this concentration, how many grams of this drug are present in 3.0 L of water?*

First, calculate the ng of the drug in 3.0 L of water using the given concentration then convert this to g.

$$3.0 \ \cancel{L} \ \times \ \frac{50 \ \cancel{\mu g}}{1 \ \cancel{L}} \ \times \ \frac{1 \times 10^{-6} \ g}{1 \ \cancel{\mu g}} \ = \ 1.5 \times 10^{-4} \ g \ \text{or} \ 2 \times 10^{-4} \ g \ \text{with 1 sig. fig.}$$

b. *Convert 50 μg /L into parts per million.*

$$\text{Parts per million} \ = \ \frac{\text{g of solute}}{\text{mL of solution}} \ \times \ 10^6$$

$$\frac{50 \ \cancel{\mu g}}{1 \ \cancel{L}} \ \times \ \frac{1 \times 10^{-6} \ g}{1 \ \cancel{\mu g}} \ \times \ \frac{1 \ \cancel{L}}{1000 \ \text{mL}} \ \times \ 10^6 \ = \ 5 \times 10^{-2} \ \text{ppm}$$

c. *Convert 50 μg /L into parts per billion.*

$$\text{Parts per billion} \ = \ \frac{\text{g of solute}}{\text{mL of solution}} \ \times \ 10^9$$

$$\frac{50 \ \cancel{\mu g}}{1 \ \cancel{L}} \ \times \ \frac{1 \times 10^{-6} \ g}{1 \ \cancel{\mu g}} \ \times \ \frac{1 \ \cancel{L}}{1000 \ \text{mL}} \ \times \ 10^9 \ = \ 5 \times 10^{1} \ \text{ppb}$$

9.101 *a. An alkyl halide is converted into the alcohol below. Write the chemical equation for this reaction.*

Alcohol A

$$+ \quad OH^- \quad \longrightarrow$$

$$+ \quad Cl^-$$

b. Draw the reaction product Alkene B.

$$\text{Alcohol A} \quad \xrightarrow[\text{heat}]{H^+} \quad \text{Alkene B}$$

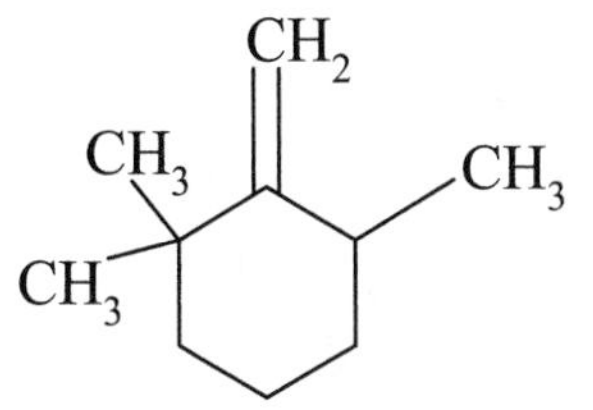

Alkene B

c. Draw the reaction product, Alcohol C.

$$\text{Alkene B} \quad + \quad H_2O \quad \xrightarrow{H^+} \quad \text{Alcohol C}$$

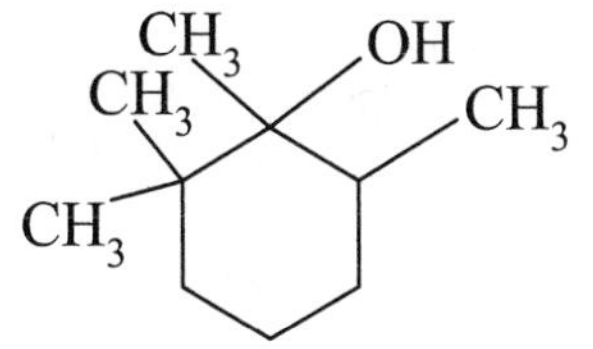

Alcohol C

d. Draw the reaction product, Alkene D.

$$\text{Alcohol C} \quad \xrightarrow[\text{heat}]{H^+} \quad \text{Alkene D}$$

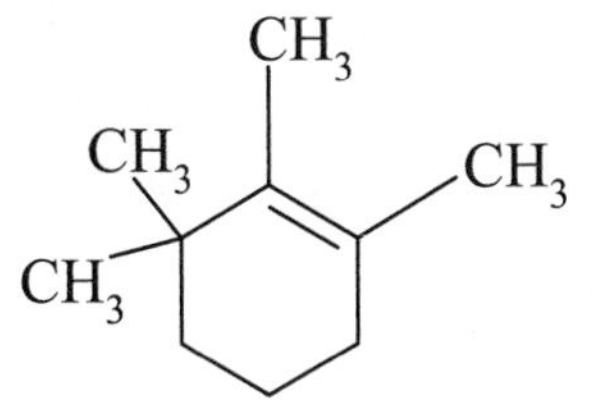

Alkene D

e. Give the IUPAC name of Alkene D.

1,2,3,3-tetramethylcyclohexene

f. Draw the reaction product.

g. Draw the reaction product.

9.103 *After reviewing the definition of the terms hemiacetal and acetal, identify any hemiacetal or acetal carbons in lactose (also known as milk sugar).*

A hemiacetal carbon atom is attached to a –OH and a –OC. An acetal carbon is attached to two –OC groups.

9.105 *Give the IUPAC name for the following compound:*

2,2,5,5-tetramethyl-3-hexanone

Chapter 10
Carbohydrates

Solutions to Problems

10.1   *a. Which pairs of glasses are mirror images?*

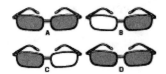

A and D are mirror images. B and C are also mirror images:

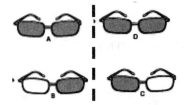

b. Which, if any, pairs of glasses can be considered to be enantiomers?

B and C can be considered enantiomers: they are mirror images of each other (see above) but they are not superimposable (see below).

10.3   *How do oligosaccharides differ from polysaccharides?*

Oligosaccharides contain from 2 to10 monosaccharide residues while polysaccharides contain more than 10.

10.5   *Classify each monosaccharide in terms of functional group and number of carbon atoms.*

$$\overset{\displaystyle O}{\underset{\displaystyle \overset{|}{OH}}{HO-CH_2-CH-\overset{||}{C}-H}}$$

a.

Aldotriose. There is an aldehyde functional group ("aldo") and there are 3 carbon atoms ("triose").

$$HO-CH_2-\underset{\overset{|}{OH}}{CH}-\overset{\overset{\displaystyle O}{||}}{C}-CH_2-OH$$

b.

Ketotetrose. There is a ketone function group ("keto") and 4 carbon atoms ("tetrose").

$$HO-CH_2-\overset{\overset{\displaystyle O}{||}}{C}-\underset{\overset{|}{OH}}{CH}-\underset{\overset{|}{OH}}{CH}-\overset{\overset{\displaystyle OH}{|}}{CH}-CH_2-OH$$

c.

Ketohexose. There is a ketone functional group and 6 carbon atoms ("hexose").

$$H-\overset{\overset{\displaystyle O}{||}}{C}-\underset{\overset{|}{OH}}{CH}-\overset{\overset{\displaystyle OH}{|}}{CH}-CH_2-OH$$

d.

Aldotetrose. There is an aldehyde functional group and 4 carbon atoms.

10.7 *Draw an example of each type of monosaccharide*

a. an aldoheptose

The name heptose indicates a seven carbon sugar. The "aldo" means it is an aldehyde. Therefore, draw a seven carbon chain with a –OH on every carbon except carbon 1 as this is where the C=O goes.

$$H-\overset{\overset{\displaystyle OH}{|}}{\underset{\underset{\displaystyle H}{|}}{C}}-\overset{\overset{\displaystyle OH}{|}}{\underset{\underset{\displaystyle H}{|}}{C}}-\overset{\overset{\displaystyle OH}{|}}{\underset{\underset{\displaystyle H}{|}}{C}}-\overset{\overset{\displaystyle OH}{|}}{\underset{\underset{\displaystyle H}{|}}{C}}-\overset{\overset{\displaystyle OH}{|}}{\underset{\underset{\displaystyle H}{|}}{C}}-\overset{\overset{\displaystyle OH}{|}}{\underset{\underset{\displaystyle H}{|}}{C}}-\overset{\overset{\displaystyle O}{||}}{C}-H$$

an aldoheptose

b. a ketononose

The name nonose indicates a nine carbon sugar. The "keto" means it is a ketone. Therefore, draw a nine carbon chain with a –OH on every carbon except for that carbon where the C=O goes.

$$H-\overset{\overset{\displaystyle OH}{|}}{\underset{\underset{\displaystyle H}{|}}{C}}-\overset{\overset{\displaystyle OH}{|}}{\underset{\underset{\displaystyle H}{|}}{C}}-\overset{\overset{\displaystyle OH}{|}}{\underset{\underset{\displaystyle H}{|}}{C}}-\overset{\overset{\displaystyle OH}{|}}{\underset{\underset{\displaystyle H}{|}}{C}}-\overset{\overset{\displaystyle OH}{|}}{\underset{\underset{\displaystyle H}{|}}{C}}-\overset{\overset{\displaystyle OH}{|}}{\underset{\underset{\displaystyle H}{|}}{C}}-\overset{\overset{\displaystyle OH}{|}}{\underset{\underset{\displaystyle H}{|}}{C}}-\overset{\overset{\displaystyle O}{||}}{C}-\overset{\overset{\displaystyle OH}{|}}{\underset{\underset{\displaystyle H}{|}}{C}}-H$$

a ketononose

10.9 *Using wedge and dashed line notation, draw the enantiomer of each molecule.*

a.

b.

An enantiomer is nonsuperimposable mirror-image of the molecule.

a.

b.

10.11 *Label the chiral carbon atom in each molecule.*

A chiral carbon atom must have four different atoms or groups of atoms attached to it.

a. 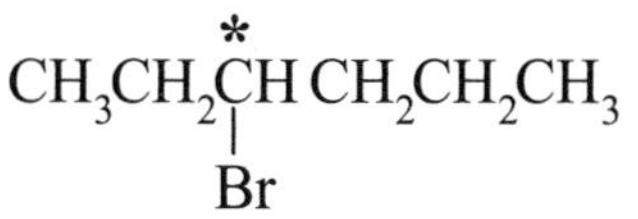   CH₃CH₂C̈HCH₂CH₂CH₃
                 |
                 Br

b. 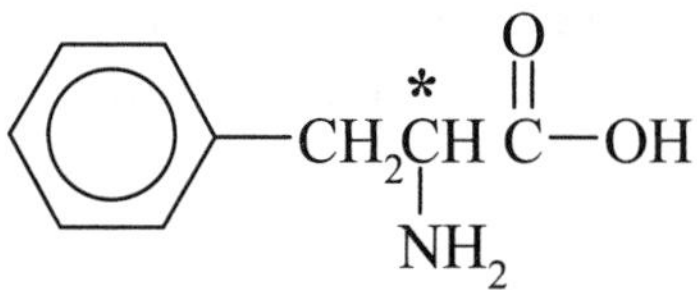   —CH₂CH C̈—OH
                        |
                        NH₂

10.13 *Draw both enantiomers of each molecule in Problem 10.11, using wedge and dashed line notation to show the three-dimensional shape about each chiral carbon atom.*

Enantiomers are molecules that are mirror images of each other and are not superimposable.

a. Note that in each structure below the propyl is out of the plane of the paper and the hydrogen points into the plane. The structures are mirror images and they are not superimposable.

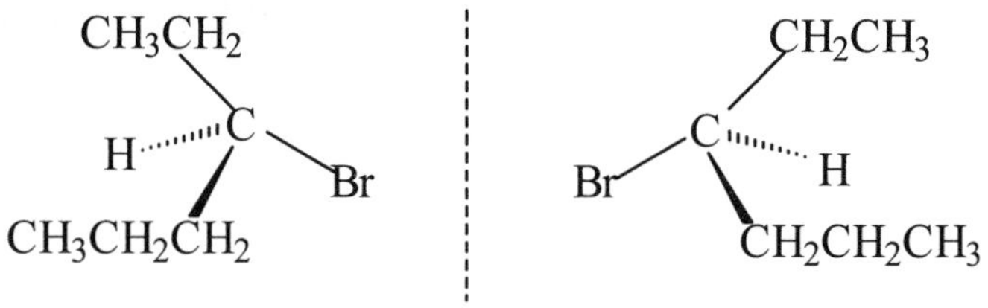

b. Note that in each structure the amino group is out of the plane of the paper and the hydrogen points into the plane. The structures are mirror images and they are not superimposable.

252

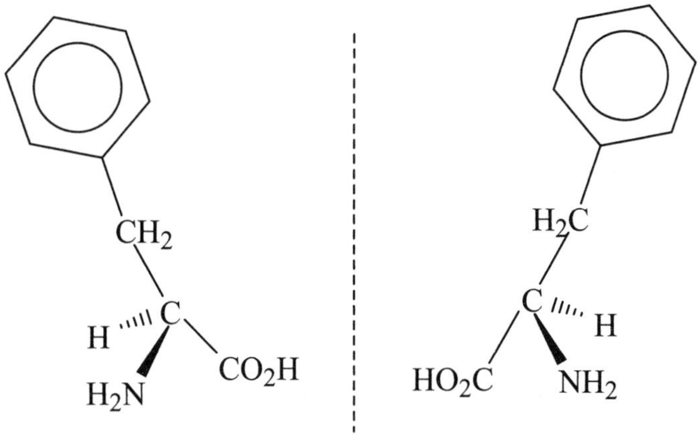

10.15 *How many chiral carbon atoms does each molecule have and how many total stereoisomers are possible?*

A chiral carbon atom has four different atoms or groups of atoms bonded to it. The chiral carbon atoms are marked by an asterisk (*) in each of the structures. The number of possible stereoisomers can be calculated using the formula 2^n, where n is the number of chiral carbon atoms.

a.

$$CH_3-\overset{*}{C}H-CH_2-\underset{*}{\overset{\overset{\displaystyle Br}{|}}{C}H}-CH_3$$
$$|$$
$$CH_2CH_3$$

$2^n = 2^2 = 4$ possible stereoisomers

b.

$$CH_3-CH-CH_2-\underset{*}{\overset{\overset{\displaystyle OH}{|}}{C}H}-CH_3$$
$$|$$
$$CH_3$$

$2^n = 2^1 = 2$ possible stereoisomers

c.

$$HO-CH_2-\underset{*}{\overset{\overset{\displaystyle OH}{|}}{C}H}-\underset{\underset{\displaystyle OH}{|}}{\overset{*}{C}H}-\overset{\overset{\displaystyle O}{||}}{C}-CH_2-OH$$

$2^n = 2^2 = 4$ possible stereoisomers

10.17 (+)-Propoxyphene is an analgesic (a painkiller).

(+)-Propoxyphene

a. Label the chiral carbon atom(s) in this molecule, using an asterisk.

b. How many other stereoisomers of propoxyphene exist? Explain.

Four stereoisomers exist for propoxyphene. The number of stereoisomers is equal to 2^n where n is the number of chiral carbons in the molecule. For propoxyphene with 2 chiral carbons, the total number of stereoisomers that exist is equal to 2^2 or 4.

c. The enantiomer of (+)-propoxyphene is an antitussive (a cough suppressant). Draw this enantiomer.

(+)-Propoxyphene   Mirror   Enantiomer

10.19 *Pick the statement that best describes the relationship between each pair of molecules: constitutional isomers, stereoisomers, different conformations of same molecule, or identical molecules.*

a.

Stereoisomers. These two molecules have the same molecular formula and the same atomic connections but different three-dimensional orientation of the groups that can be interchanged only by breaking bonds.

$$\text{H}—\overset{\overset{\textstyle H}{|}}{\underset{\underset{\textstyle H}{}}{C}}—CH_3 \qquad CH_3—\overset{\overset{\textstyle H}{|}}{C}—H \ (H)$$

b.

Different conformations of the same molecules/ identical molecules. Flipping one of them horizontally by 180° (or rotating the molecule) will produce a molecule identical to the other.

$$\overset{CH_3}{\underset{H}{>}}C=C\overset{CH_3}{\underset{H}{<}} \qquad \overset{CH_3}{\underset{H}{>}}C=C\overset{H}{\underset{CH_3}{<}}$$

c.

Stereoisomers. These two molecules have the same molecular formula and the same atomic connections but different three-dimensional orientation of the groups that can be interchanged only by breaking bonds.

$$\overset{CH_2—Br}{\underset{\underset{\textstyle CH_3}{}}{\overset{|}{C}}}\!\!\!\!H\; CH_3 \qquad \overset{CH_3}{\underset{}{\overset{|}{C}}}\!\!\!\!H\; CH_2—CH_3,\; Br$$

d.

Constitutional isomers. These 2 molecules have the same molecular formulas but the atomic connections are different.

10.21 *a. Using Fischer projections, draw the two enantiomers of the monosaccharide.*
b. Label the molecules in your answer to part a as being D or L.

$$HO—CH_2—\underset{\underset{\textstyle OH}{|}}{CH}—\overset{\overset{\textstyle O}{\|}}{C}—H$$

a. and *b.*

In drawing a Fischer projection for a monosaccharide, the carbon atoms run vertically with the aldehyde or ketone group at or near the top. In D sugars, the –OH bonded to the chiral carbon atom farthest from the C=O is pointed to the right. In L sugars, this same –OH group is pointed to the left.

$$
\begin{array}{ccc}
\underset{\parallel}{\overset{O}{C}}-H & & \underset{\parallel}{\overset{O}{C}}-H \\
H-\!\!\!-\!\!\!-OH & & HO-\!\!\!-\!\!\!-H \\
CH_2OH & & CH_2OH \\
D & & L
\end{array}
$$

c. Classify the monosaccharide in part a in terms of functional group and number of carbon atoms.

Aldotriose. There is an aldehyde group and 3 carbon atoms.

10.23 *a. How many chiral carbon atoms does D-lyxose contain?*

$$
\begin{array}{c}
\underset{\parallel}{\overset{O}{C}}-H \\
HO-\!\!\!-\!\!\!-H \\
HO-\!\!\!-\!\!\!-H \\
H-\!\!\!-\!\!\!-OH \\
CH_2OH \\
\end{array}
$$

D-Lyxose

3 chiral carbon atoms, as marked by asterisks below.

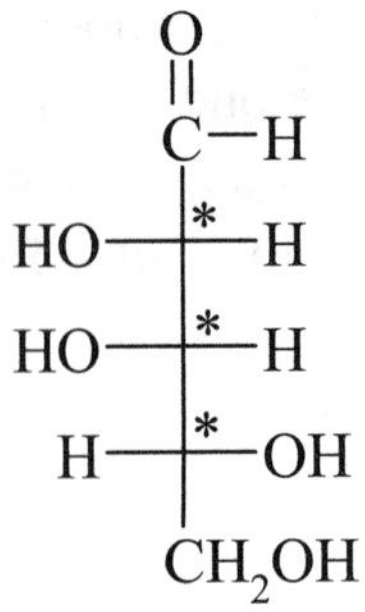

$$\overset{\displaystyle O}{\underset{\displaystyle C-H}{\|}}$$

```
          O
          ‖
          C—H
HO ——*—— H
HO ——*—— H
 H ——*—— OH
       CH₂OH
```

D-Lyxose

b. How many total stereoisomers are possible for this aldopentose?

8 possible stereoisomers: $2^3 = 8$.

c. Draw and name the enantiomer of D-lyxose.

The enantiomer of D-lyxose is obtained by drawing the mirror image of D-lyxose.

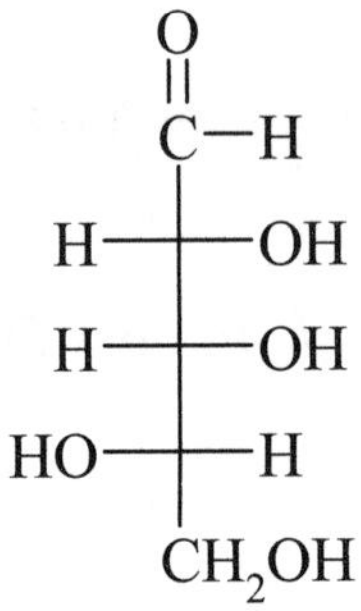

```
          O
          ‖
          C—H
 H ———— OH
 H ———— OH
HO ———— H
       CH₂OH
```

L-Lyxose

d. D-Ribose is a diastereomer of D-lyxose. Draw two additional D-diastereomers of this monosaccharide.

Diastereomers are stereoisomers that are not mirror images. Move one or more of the —OH groups in directions unlike those found in D-lyxose. Be sure that you haven't drawn L-lyxose (part c) and that you have drawn D sugars.

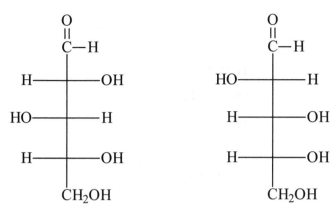

D-diastereomers

10.25 *Pick the statement that describes the relationship between each pair of molecules; enantiomers, diastereomers, constitutional isomers, molecules that are neither stereoisomers nor constitutional isomers.*

a. D-ribose and L-ribose

Enantiomers. These two molecules are mirror-images of each other as indicated by the D- and L- difference in orientation.

b. D-glucose and D-galactose

Diastereomers. These two are stereoisomers that are not mirror-images of each other.

c. L-glucose and D-galactose

Diastereomers. These two are stereoisomers that are not mirror-images of each other.

d. D-glucose and D-2-deoxyribose

Molecules that are not stereoisomers or constitutional isomers. These two molecules do not have the same molecular formula.

10.27 *Draw each of the following glucose derivatives.*

a. D-2-deoxyglucose

The –OH group at carbon number 2 is replaced ("2-deoxy") by a H atom.

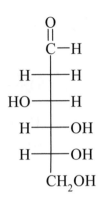

b. *D-glucosamine*

An –OH group is replaced by an amine group.

$$
\begin{array}{c}
O \\
\parallel \\
C-H \\
H-\!\!-NH_2 \\
HO-\!\!-H \\
H-\!\!-OH \\
H-\!\!-OH \\
CH_2OH
\end{array}
$$

c. *D-glucuronic acid*

The terminal –CH₂OH group is replaced by a –COOH group.

$$
\begin{array}{c}
O \\
\parallel \\
C-H \\
H-\!\!-OH \\
HO-\!\!-H \\
H-\!\!-OH \\
H-\!\!-OH \\
C-OH \\
\parallel \\
O
\end{array}
$$

d. *sorbitol*

The aldehyde group (-CHO) at carbon 1 is replaced by a –CH₂OH group.

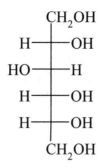

10.29 *For the molecules in Problem 10.27,*
 a. To which class of monosaccharide derivative does each belong?
 b. How many chiral carbon atoms does each have?
 c. How many stereoisomers are possible for each?

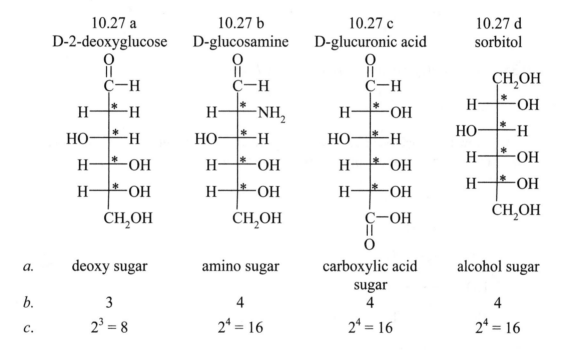

	10.27 a D-2-deoxyglucose	10.27 b D-glucosamine	10.27 c D-glucuronic acid	10.27 d sorbitol
a.	deoxy sugar	amino sugar	carboxylic acid sugar	alcohol sugar
b.	3	4	4	4
c.	$2^3 = 8$	$2^4 = 16$	$2^4 = 16$	$2^4 = 16$

10.31 *When D-gluconic acid is produced from D-glucose, which functional group is oxidized?*

The aldehyde on carbon 1.

10.33 *a. Draw all possible aldotetroses.*
 b. Which are D and which are L sugars?

The name tetrose indicates a four carbon sugar. The "aldo" means it is an aldehyde. Therefore, draw a four carbon chain with a –OH on every carbon except carbon 1, as this is where the C=O goes. To make all possible aldotetroses, arrange the –OH groups of carbons 2 and 3 in as many different ways as possible. (Note: Two chiral carbons tell you there are $2^2 = 4$ different structures possible). D structures have the –OH group on the right at carbon 3, while L structures have the –OH group on the left.

```
     O                    O                    O                    O
     ||                   ||                   ||                   ||
     C—H                  C—H                  C—H                  C—H
HO——+——H            H——+——OH            H——+——OH          HO——+——H
 H——+——OH           H——+——OH           HO——+——H           HO——+——H
    CH₂OH               CH₂OH                CH₂OH                CH₂OH

     D                    D                    L                    L
```

10.35   *Draw each galactose derivative (see Figure 10.10).*

   a. D-2-deoxygalactose

Draw the galactose as shown in Figure 10.10 and then remove the –OH group from carbon 2 and replace it with hydrogen.

```
        O
        ||
        C—H
  H——+——H
 HO——+——H
 HO——+——H
  H——+——OH
       CH₂OH
```

   b. galactitol (the alcohol sugar derived from galactose)

Draw galactose as shown in Figure 10.10. Convert the aldehyde group into an alcohol.

261

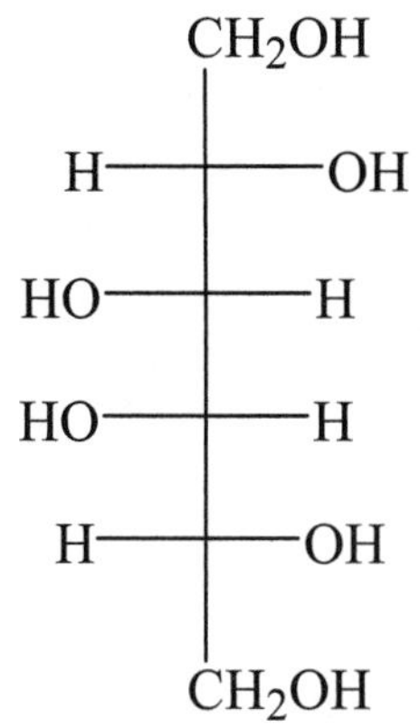

CH_2OH

H——OH

HO——H

HO——H

H——OH

CH_2OH

c. D-galacturonic acid (carbon atom # 6 of galactose is oxidized to a carboxylic acid)

Draw galactose as shown in Figure 10.10. Convert the alcohol group on carbon 6 to a carboxylic acid.

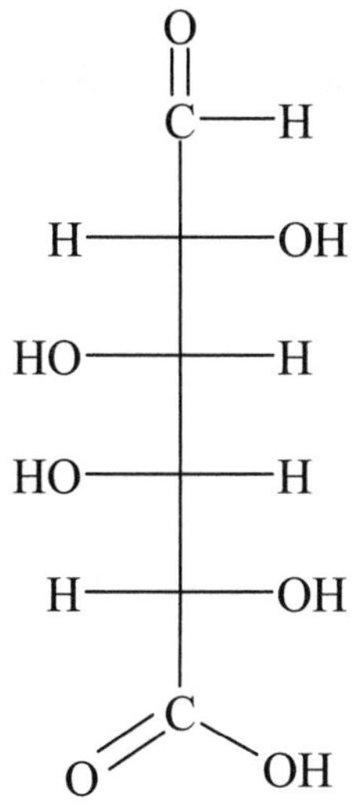

O
||
C——H

H——OH

HO——H

HO——H

H——OH

O=C—OH

10.37 *Draw the product obtained when D-talose*

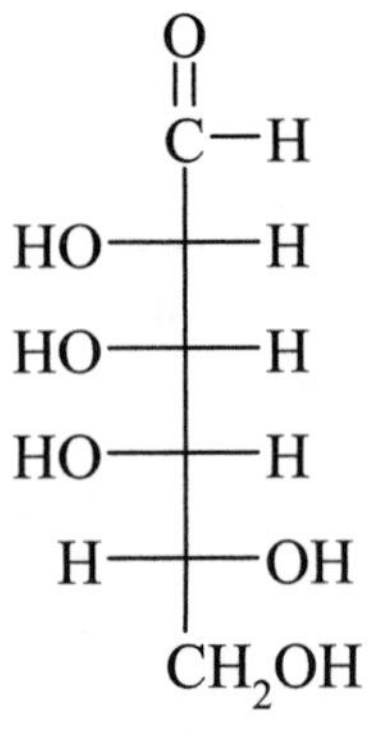

$$
\begin{array}{c}
\overset{\displaystyle O}{\underset{\displaystyle \|}{}} \\
C-H \\
HO-\!\!\!\!-H \\
HO-\!\!\!\!-H \\
HO-\!\!\!\!-H \\
H-\!\!\!\!-OH \\
CH_2OH
\end{array}
$$

D-Talose

a. reacts with H_2 and Pt.

When a monosaccharide is reduced, the carbon-oxygen double bond of the aldehyde or ketone group is reduced to an alcohol:

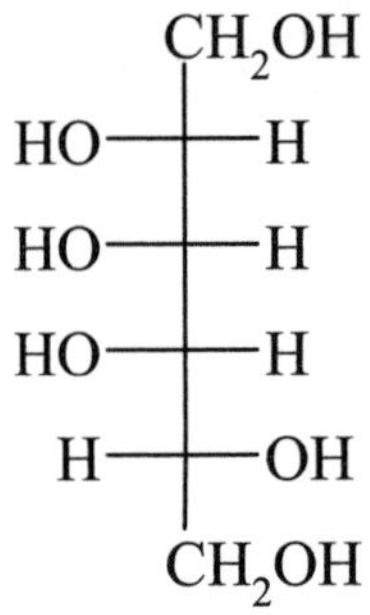

$$
\begin{array}{c}
CH_2OH \\
HO-\!\!\!\!-H \\
HO-\!\!\!\!-H \\
HO-\!\!\!\!-H \\
H-\!\!\!\!-OH \\
CH_2OH
\end{array}
$$

reduced form of D-talose

b. reacts with Benedict's reagent.

Oxidizing carbon 1 of D-talose produces a carboxylic acid group at carbon 1:

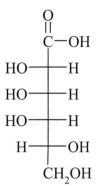

oxidized form of D-talose at carbon 1

10.39 *Define the term reducing sugar.*

A reducing sugar is a carbohydrate that gives a positive Benedict's test (in the process of being oxidized, the sugar reduces Cu^{2+} ion present in the reagent).

10.41 *a. Draw the D-aldohexose that gives the following alcohol sugar, when treated with H_2 and Pt.*

```
        CH₂OH
   H ───┼─── OH
  HO ───┼─── H
  HO ───┼─── H
   H ───┼─── OH
        CH₂OH
```

H_2 and Pt will reduce an aldehyde to an alcohol. Start with the structure shown in the problem. Convert the alcohol into an aldehyde.

```
         O
         ‖
         C ── H
   H ───┼─── OH
  HO ───┼─── H
  HO ───┼─── H
   H ───┼─── OH
         CH₂OH
```

b. Is the aldohexose a reducing sugar?

Yes. The aldehyde group can be oxidized by Cu^{2+}.

264

c. Is the aldohexose a deoxy sugar?

No. Deoxy means that an –OH group has been replaced by hydrogen. This is not the case for the D-aldohexose.

d. Is the aldohexose an amino sugar?

No. Amino sugar means that an –OH has been replaced by –NH_2. This is not the case for the D-aldohexose.

e. Draw the L- aldohexose that also gives the alcohol sugar above, when treated with H_2 and Pt.

```
        O
        ||
        C—H
 HO ———— H
  H ———— OH
  H ———— OH
 HO ———— H
      CH₂OH
```

10.43   *How does an α anomer differ from a β anomer?*

As pyranoses and furanoses are typically drawn (Figures 10.14 and 10.15), the hemiacetal -OH points down in an α anomer and up in a β anomer.

10.45   Draw each molecule.

a. β-D-glucopyranose

```
        CH₂OH
          |
 H        |————O
  \       H        OH
   \             /
    \  OH    H  /
 HO          H
    |        |
    H        OH
```

b. α-D-ribofuranose

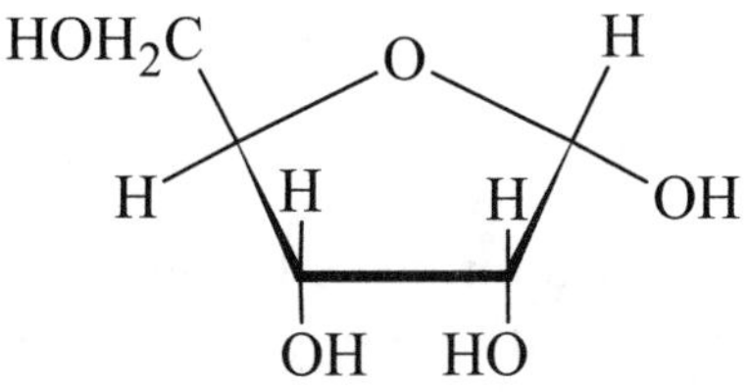

c. β-D-galactopyranose

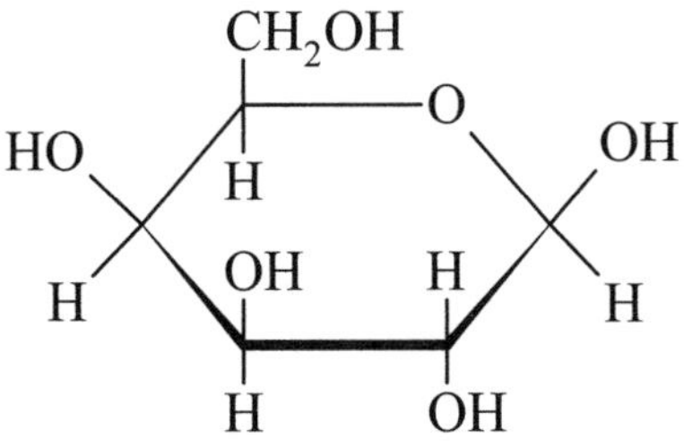

d. α- D-arabinopyranose (see Practice Problem 10.6)

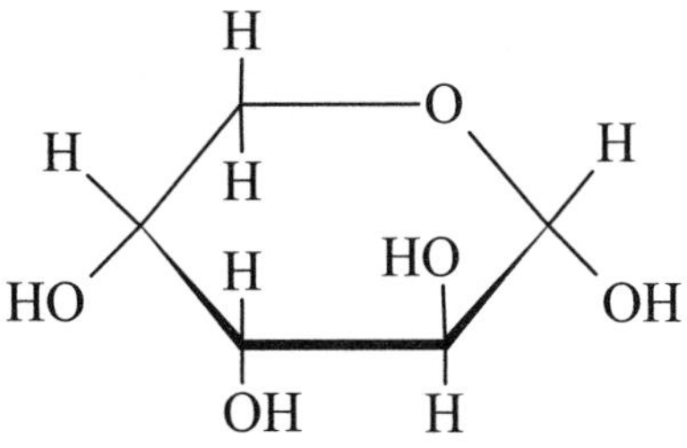

10.47 *Each molecule in Problem 10.45 has how many chiral carbon atoms and is one of how many total stereoisomers?*

Molecule	Number of Chiral Carbons	Total Number of Stereoisomers
β-D-glucopyranose	5	32
α -D-ribofuranose	4	16
β-D-galactopyranose	5	32

10.49   *D-lyxose (see Problem 10.23) is a diastereomer of D-ribose. Draw α-D-lyxofuranose.*

A lyxofuranose molecule is a lyxose molecule that has formed a five-membered cyclic hemiacetal. The five-membered ring is drawn with the oxygen at the back and the hemiacetal carbon on the right side. The α-notation indicates that the –OH on the hemiacetal carbon is pointing down.

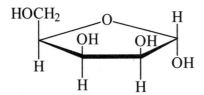

10.51   *Draw β-D-mannopyranose (see Figure 10.12).*

10.53   *a. Draw the α pyranose anomer of D-2-deoxyglucose.*

D-2-deoxyglucose   α pyranose anomer

b. Draw the β pyranose anomer of D-glucuronic acid.

D-glucuronic   β pyranose anomer

10.55 *Define the term mutarotation.*

Mutarotation is the process of converting back and forth from an α anomer to the open form to the β anomer.

10.57 *Cellobiose is a reducing sugar. Write a reaction equation that shows why.*

When cellobiose ring opens, it contains an aldehyde that can be oxidized by Cu^{2+}.

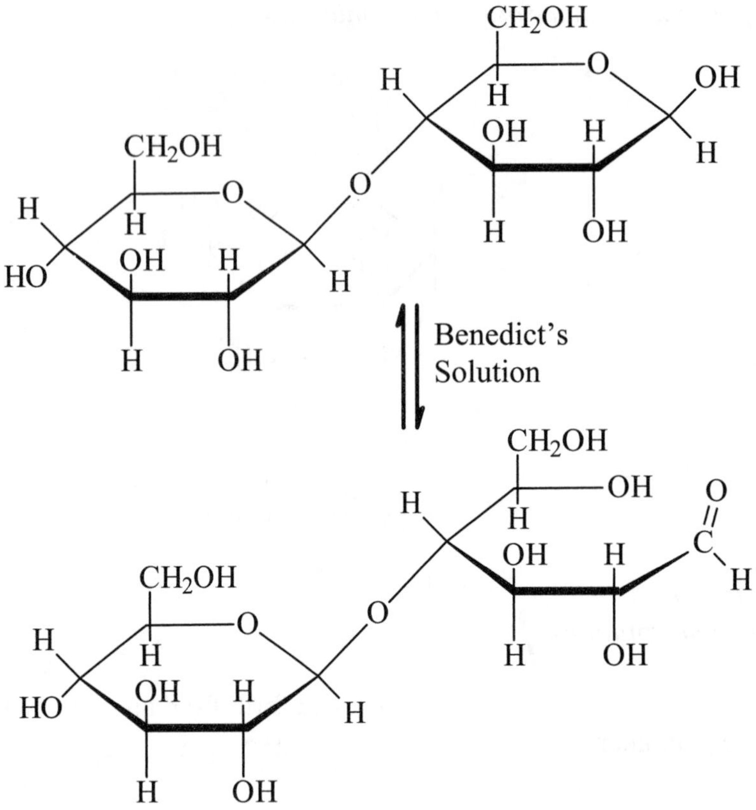

CH₂OH

H   H   O   OH

OH   H

H   OH

CH₂OH

H   O

H   O

HO   OH   H

H

H   OH

Benedict's
Solution

CH₂OH

H   OH   O

H   C

OH   H   H

H   OH

CH₂OH

H   O   O

HO   OH   H

H

H   OH

10.59 *Gentiobiose consists of two D-glucose residues joined by a β-(1 →6) glycosidic bond.*

a. Draw this disaccharide.

Draw two β-D-glucopyranose molecules. From one of them, remove the hydrogen from the –OH on the hemiacetal carbon (carbon 1). From the other remove the –OH from carbon 6. Draw a bond connecting the two molecules.

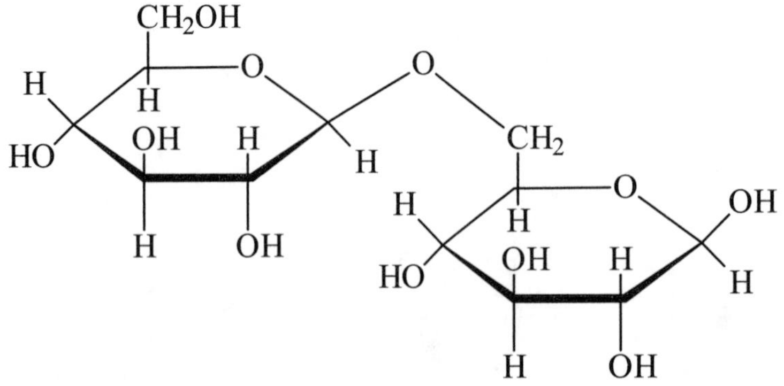

CH₂OH

H   O   O

HO   H

OH   H   H

CH₂

H   OH

H   O

HO   OH   H   OH

H

H   OH

b. Is this gentiobiose a reducing sugar? Explain.

Yes. When the β hemiacetal (at the right side of the molecule) undergoes mutarotation, the resulting aldehyde group can be oxidized by Cu^{2+}.

10.61   *Vanillin β-D-glucoside gives vanilla extract its flavor. Draw the two products obtained when this acetal is hydrolyzed.*

Vanillin β-D-Glucoside

Hydrolysis of an acetal bond produces a hemiacetal and an alcohol.

10.63   *Draw a disaccharide consisting of two D-glucose residues that is not a reducing sugar.*

The disaccharide can be made by linking two D-glucose molecules with an α, β-$(1 \longleftrightarrow 1)$ glycosidic bond. The resulting structure contains no hemiacetal group. Therefore, it is unable to mutarotate, and it is not a reducing sugar.

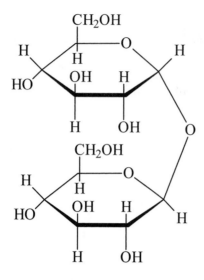

10.65 *a. Is trehalose, a disaccharide found in a wide range of living things, a reducing sugar?*

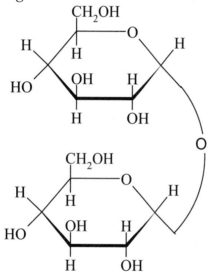

Trehalose

No. The molecule contains no hemiacetal groups.

b. Which describes the glycosidic bond in this disaccharide: α,α-(1↔1),
α-(1→2), α-(1→3), α-(1→4), α-(1→5), or α-(1→6)?

α,α-(*1↔1*)
The glycosidic bond involves the hemiacetal carbon (carbon 1) of each ring, each of which is the α anomer.

271

10.67 *Draw the product obtained when lactose is reacted with Benedict's reagent.*

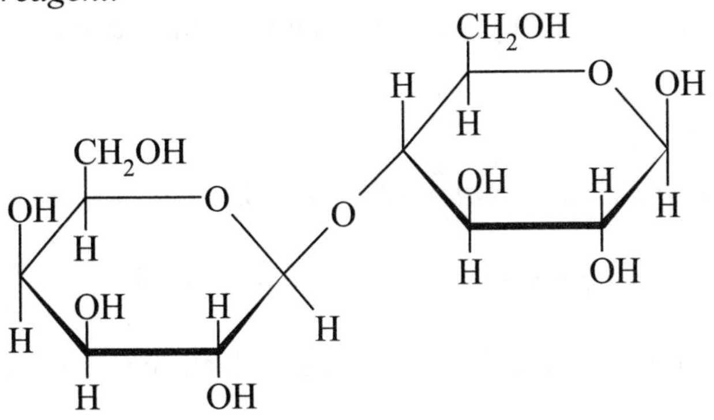

lactose (β anomer)

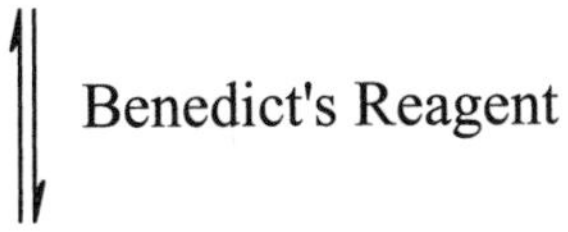

Benedict's Reagent

The carbonyl group of the open form of the monosaccharide residue has been oxidized into a carboxyl group.

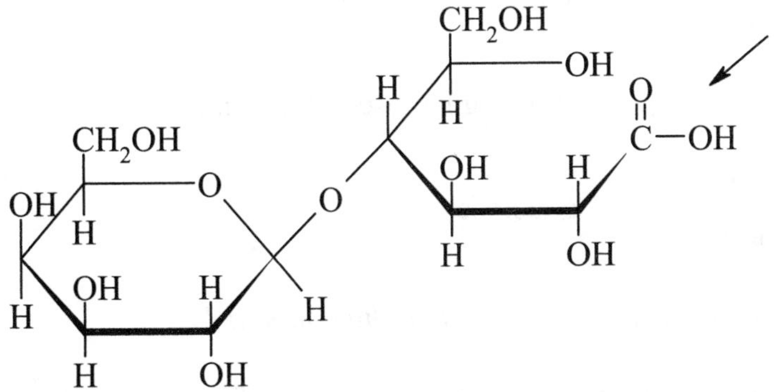

lactose (open-ring)

10.69 *True or False?*

a. Lactose and cellobiose are stereoisomers.

True. Lactose and cellobiose have the same molecular formula and atomic connections but have different three-dimensional shapes that are interchanged only by breaking bonds.

b. Lactose and cellobiose are enantiomers.

False. They are not enantiomers because they are not mirror images of each other.

c. Lactose and cellobiose are diastereomers.

True. They are diastereomers because they are stereoisomers that are not enantiomers.

10.71   *Human milk contains lactose at a concentration of 7% (w/v).*

    a. How many grams of lactose are present in 15 mL of human milk?

Use the given concentration above to calculate the grams of lactose remembering that

$$\% \text{ w} / \text{v} \; = \; \frac{\text{g solute}}{\text{mL solution}} \; \text{x } 100\%$$

In 15 mL of human milk,

$$7\% \text{ w} / \text{v} \; = \; \frac{7 \text{ g solute}}{100 \text{ mL solution}} \; \text{x } 100\%$$

Therefore, a 7% (w/v) solution contains 7 g lactose/100 mL milk.

$$15 \; \cancel{\text{mL}} \text{ milk } \; \text{x} \; \frac{7 \text{ g lactose}}{100 \; \cancel{\text{mL}} \text{ milk}} \; = \; 1 \text{ g lactose}$$

    b. What is the molar concentration of lactose in human milk?

First, convert 100 mL to 0.100 L then use the molar mass of lactose (342 g/mol) to calculate its molar concentration in milk.

$$\left(7 \; \cancel{\text{g}} \text{ lactose } \; \text{x} \; \frac{1 \text{ mol lactose}}{342 \; \cancel{\text{g}} \text{ lactose}} \right) \div \; 0.100 \text{ L } = \; 0.2 \text{ M}$$

    c. What is the parts per million concentration of lactose in human milk?

Recall the formula for calculating parts per million:

$$\text{Parts per million } = \; \frac{\text{g of solute}}{\text{mL of solution}} \; \text{x } 10^6$$

This formula gives

$$\frac{7 \text{ g lactose}}{100 \text{ mL milk}} \times 10^6 = 7 \times 10^4 \text{ ppm}$$

10.73 *a. How is the structure of amylose similar to that of amylopectin?*

Amylose and amylopectin each consist of glucose residues connected by
α–(1→4) glycosidic bonds.

b. How is the structure of amylase different from that of amylopectin?

Amylose contains only α–(1→4) glycosidic bonds while amylopectin contains
α–(1→4) glycosidic bonds and α– (1→6) glycosidic bonds at the branching
points.

10.75 *How is the structure of glycogen similar to that of amylopectin?*

Glycogen and amylopectin each consist of glucose residues joined by α-(1→4)
and α-(1→6) glycosidic bonds.

10.77 *Why can humans use starch as a food source, but not cellulose?*

Humans have the enzymes necessary for hydrolyzing and digesting starch
(amylase, maltase, and debranching enzymes) but not the enzymes required
for hydrolyzing the type of glycosidic bonds found in cellulose.

10.79 *A particular lichen produces the polysaccharide below.*

a. Name the monosaccharide residues A-C and specify whether each is a furanose or a pyranose.

A, B, and C are D-glucose residues (pyranose form).

b. Label [β(1 → 4), etc.] the glycosidic bonds 1 and 2.

Both glycosidic bonds 1 and 2 are β-(1→6).

c. Is the molecule a homopolysaccharide or is it a heteropolysaccharide?

Homopolysaccharide. The monosaccharide residues are the same.

10.81 *What is invert sugar and why is it used in foods?*

Invert sugar is the sweetening mixture produced when sucrose is hydrolyzed into glucose and fructose. It is used in foods because it is sweeter than sucrose and it is cost-effective.

10.83 *The structure of sucralose is closest to which of the four disaccharides discussed in Section 10.6?*

Sucralose is a derivative of sucrose, a naturally occurring oligosaccharide.

10.85 *What is lactose intolerance and what are some of the options for dealing with it?*

Lactose intolerance is a deficiency in β-galactosidase, the enzyme that catalyzes the hydrolysis of the β-(1→4) glycosidic bond in lactose. Persons with this disorder can deal with it by avoiding dairy products, by using dairy products from which lactose has been removed, and by taking tablets containing β-galactosidase.

10.87 *a. Name each monosaccharide residue in rebaudioside A (Figure 10.25) and specify whether each is a furanose or a pyranose.*

All monosaccharide residues are D-glucose residues (pyranose). See figure below for the structure of rebaudioside A.

b. Identify [β-(1→4), etc.] each glycosidic bond in rebaudioside A.

β-(1→3) and β-(1→2)

c. Is this compound a reducing sugar?

No, it is not a reducing sugar. The molecule contains no hemiacetal groups.

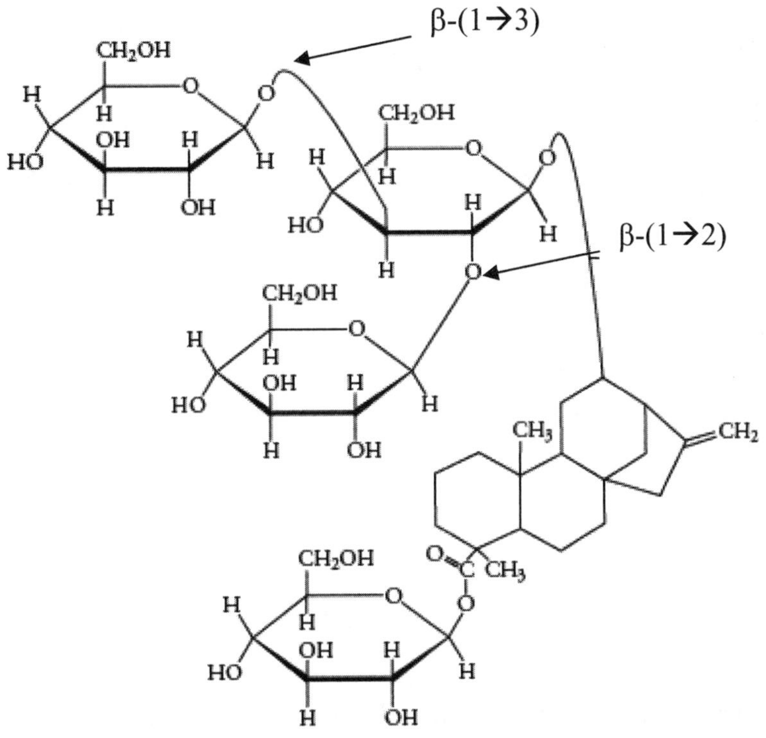

Rebaudioside A

276

10.89　*a. Draw a Fischer projection of D-2-deoxyribose.*

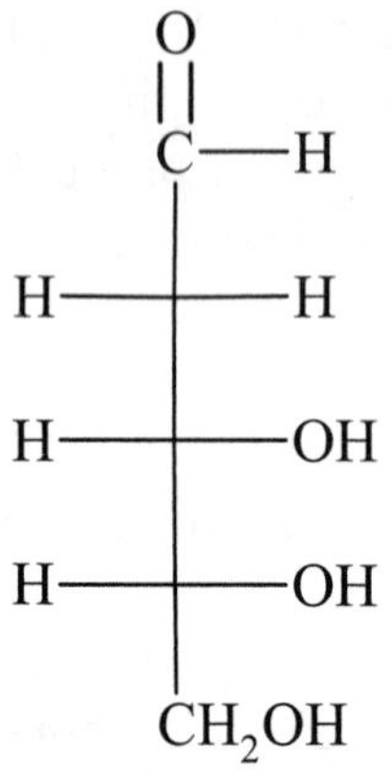

```
        O
        ‖
        C—H
        |
  H ————+———— H
        |
  H ————+———— OH
        |
  H ————+———— OH
        |
      CH₂OH
```

b. Draw a diastereomer of D-2-deoxyribose that is a D-sugar.

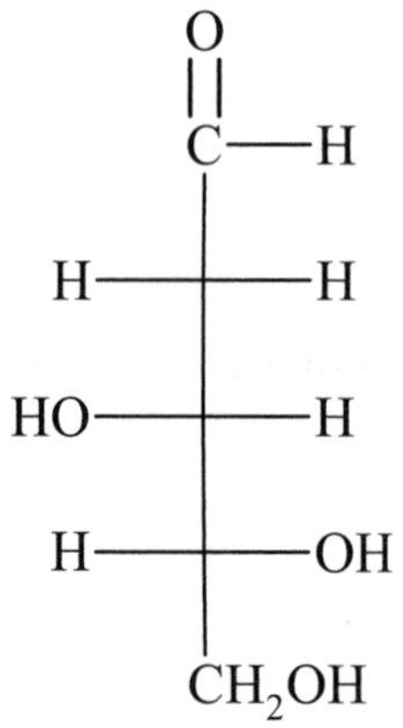

```
        O
        ‖
        C—H
        |
  H ————+———— H
        |
  HO ———+———— H
        |
  H ————+———— OH
        |
      CH₂OH
```

c. Draw L-2 deoxyribose.

```
        O
        ‖
        C—H
        |
  H ————+———— H
        |
  HO ———+———— H
        |
  HO ———+———— H
        |
      CH₂OH
```

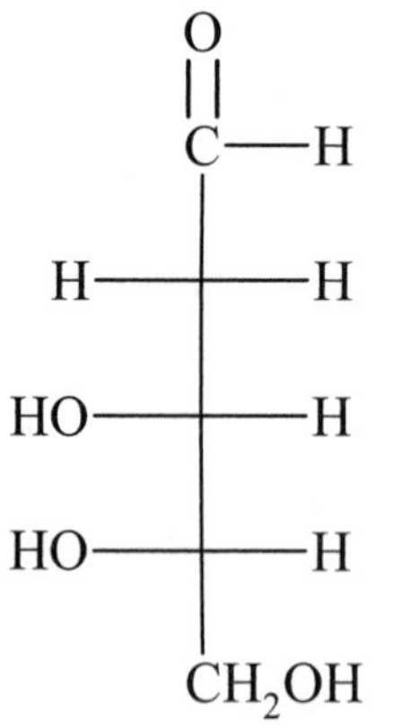

d. Draw a diastereomer of D-2-deoxyribose that is an L-sugar, but is not L-2-deoxyribose.

277

e. *Which, if any of the molecules in parts a-d are pairs of enantiomers?*

a. and c., b. and d.

f. *Is D-2-deoxyribose a reducing sugar? Why or why not?*

Yes. D-2-Deoxyribose has an aldehyde group that can be oxidized by Cu^{2+}.

g. *Draw α-D-2-deoxyribofuranose.*

h. *Draw β-D-2- deoxyribofuranose.*

i. *Are the molecules in parts g and h enantiomers, diastereomers, or neither?*

diastereomers

j. *Is α-D-2-deoxyribofuranose a reducing sugar? Why or Why not?*

Yes. When the α hemiacetal group undergoes mutarotation, the resulting aldehyde group can be oxidized by Cu^{2+}.

k. Draw a disaccharide that consists of two β-D-2- deoxyribofuranose residues joined by a β(1→5) glycosidic bond.

l. Is the disaccharide a reducing sugar? Why or why not?

Yes. When the β hemiacetal group (at the right side of the disaccharide molecule) undergoes mutarotation, the resulting aldehyde group can be oxidized to Cu^{2+}.

Chapter 11
Lipids and Membranes

Solutions to Problems

11.1   *In Chapter 8 we saw that, in the presence of H^+, a carboxylic acid and an alcohol will react to form an ester plus water.*

$$CH_3\text{-}\overset{\overset{\displaystyle O}{\|}}{C}\text{-}OH \quad + \quad HOCH_3 \quad \underset{}{\overset{H^+}{\rightleftharpoons}} \quad CH_3\text{-}\overset{\overset{\displaystyle O}{\|}}{C}\text{-}O\text{-}CH_3 \quad + \quad H_2O$$

Triglycerides are formed in an enzyme-catalyzed version of this same reaction.

a. For the reaction below, circle the reactant atoms that combine to form the water molecules.

b. Draw the triglyceride product by combining the remaining (not circled) parts of the reactant molecules.

11.3   *Why is the melting point of lauric acid (Table 11.1) lower than that of myristic acid?*

Lauric acid has a shorter hydrocarbon tail than myristic acid, so London force interactions between lauric acid molecules are weaker.

11.5   *Draw line-bond structures of the following (see Table 11.1).*

  a. *lauric acid.*

  b. *linolenic acid.*

11.7   *Sodium palmitate, $CH_3(CH_2)_{14}CO_2Na$, has a higher melting point than palmitic acid, $CH_3(CH_2)_{14}CO_2H$. Why?*

The ionic bonds that hold the ions in sodium palmitate to one another are stronger than the noncovalent interactions that hold one palmitic acid molecule to another.

11.9   *In terms of structure, what distinguishes fatty acids from the carboxylic acids that were discussed in Chapter 8?*

Fatty acids are carboxylic acids that typically have between 12 and 20 carbon atoms.

11.11 *Fatty acids are carboxylic acids that typically contain 12-20 carbon atoms. Some biological molecules contain carboxylic acid residues longer than 20 carbon atoms. Would these fatty acids with more than 20 carbon atoms in length be considered lipids? Explain.*

Lipids are defined as biochemical compounds that are water insoluble. Fatty acids longer than 20 carbons follow this definition and therefore are considered lipids.

11.13   *Draw skeletal structures for the products formed when the beeswax ester (Table 11.2) is saponified (hydrolyzed).*

Saponification is the hydrolysis of the ester bonds found in waxes.

and

11.15   *List the biological functions of waxes.*

Some of the biological functions of waxes are to form a protective layer for retaining water or keeping water out of an organism and to act as a form of energy storage.

11.17   *The fragrance of spermaceti, a wax produced by whales, once made it important to the perfume industry. One of the main constituents of spermaceti is cetyl palmitate, which is formed from palmitic acid and cetyl alcohol, $CH_3(CH_2)_{14}CH_2OH$. Draw the condensed structure of cetyl palmitate.*

First find the formula for palmitic acid, $CH_3(CH_2)_{14}CO_2H$ in Table 11.1. Next remember that the reaction between a carboxylic acid and an alcohol forms an ester. Remove the –OH from the acid and the –H from the alcohol and join the two ends together.

$$CH_3(CH_2)_{14}\overset{\displaystyle O}{\overset{\|}{C}}-OCH_2(CH_2)_{14}CH_3$$

11.19   *a. Draw skeletal structures for the fatty acid and alcohol from which the following king fisher green wax ester is made.*

$$CH_3CH_2CH_2CHCH_2CHCH_2CHCH_2CHC-OCH_2CHCH_2CHCH_2CHCH_2CHCH_2CH_2CH_3$$

(with CH₃ substituents below)

The ester is formed from the following alcohol and carboxylic (fatty) acid:

and

b. Write a chemical equation that shows how the alcohol in your answer to part a. could be converted into the carboxylic acid in your answer to part a.

$\xrightarrow{K_2Cr_2O_7}$

11.21 *Some of the ester molecules present in beeswax have an –OH group in the ω-3 position of the alcohol residue. Draw a possible structure for one of these esters. Refer to Health Link: Omega-3 Fatty Acids.*

$$CH_3(CH_2)_{14}-\overset{\displaystyle O}{\overset{\displaystyle \|}{C}}-O-CH_2(CH_2)_{26}\overset{|}{\underset{OH}{C}}H-CH_2CH_3 \quad \text{or} \quad CH_3(CH_2)_{14}CO_2CH_2(CH_2)_{26}CH(OH)CH_2CH$$

ω-3 carbon   ω-3 carbon

11.23 *List four biological functions of triglycerides.*

Energy source, thermal insulation, padding, and buoyancy.

11.25 *Write a balanced reaction equation for the complete hydrogenation of palmitoleic acid (Table 11.1).*

Hydrogenation is the process of adding H_2 to carbon atoms of a double bond.

$$CH_3(CH_2)_5CH\!\!=\!\!CH(CH_2)_7CO_2H \quad \xrightarrow[\text{Pt}]{H_2} \quad CH_3(CH_2)_5CH_2CH_2(CH_2)_7CO_2H$$

11.27 *Draw a triglyceride made from glycerol, myristic acid, palmitic acid, and oleic acid. Would you expect this triglyceride to be a liquid or a solid at room temperature? Explain.*

Triglycerides are composed of three fatty acid residues and a glycerol residue. Look up the formulas for each of the fatty acids and attach them to the glycerol molecule in the same way that you drew the esters in the previous problems.

$$CH_3(CH_2)_{12}\overset{\displaystyle O}{\overset{\|}{C}}\!-\!O\!-\!CH_2$$

$$CH_3(CH_2)_{14}\overset{\displaystyle O}{\overset{\|}{C}}\!-\!O\!-\!CH$$

$$CH_3(CH_2)_7CH\!\!=\!\!CH(CH_2)_7\overset{\displaystyle O}{\overset{\|}{C}}\!-\!O\!-\!CH_2$$

This triglyceride contains more saturated than unsaturated fatty acid residues, so it should be a solid at room temperature.

11.29 *a. Draw the products formed when the triglyceride is saponified.*

$$CH_3(CH_2)_{14}\overset{\displaystyle O}{\overset{\|}{C}}\!-\!O\!-\!CH_2$$

$$CH_3CH_2(CH\!\!=\!\!CHCH_2)_3(CH_2)_6\overset{\displaystyle O}{\overset{\|}{C}}\!-\!O\!-\!CH$$

$$CH_3(CH_2)_7CH\!\!=\!\!CH(CH_2)_7\overset{\displaystyle O}{\overset{\|}{C}}\!-\!O\!-\!CH_2$$

Saponification is another term for ester hydrolysis in the presence of OH⁻. To show the products, take the molecule apart the same as any other ester. The products will be the three fatty acid anions and glycerol.

$$CH_3(CH_2)_{14}\overset{\overset{\displaystyle O}{\|}}{C}-O^-$$ $$HO-CH_2$$

$$CH_3CH_2(CH=CHCH_2)_3(CH_2)_6\overset{\overset{\displaystyle O}{\|}}{C}-O^-$$ $$HO-\underset{|}{C}H$$

$$CH_3(CH_2)_7CH=CH(CH_2)_7\overset{\overset{\displaystyle O}{\|}}{C}-O^-$$ $$HO-CH_2$$

b. For each product from part a., state whether it is hydrophobic, hydrophilic, or amphipathic.

$$CH_3(CH_2)_{14}\overset{\overset{\displaystyle O}{\|}}{C}-O^-$$

amphipathic

$$CH_3CH_2(CH=CHCH_2)_3(CH_2)_6\overset{\overset{\displaystyle O}{\|}}{C}-O^-$$

amphipathic

$$HO-CH_2$$

$$HO-CH$$

$$CH_3(CH_2)_7CH=CH(CH_2)_7\overset{\overset{\displaystyle O}{\|}}{C}-O^-$$

amphipathic

$$HO-CH_2$$

hydrophilic

11.31 *Vegetable oils tend to become rancid more rapidly than do animal fats. Why?*

Vegetable oils have more carbon-carbon double bonds than animal fats. It is the oxidation of these bonds that causes oils to become rancid.

11.33 *What does it mean to partially hydrogenate vegetable oil?*

Partial hydrogenation of vegetable oil involves the addition of hydrogen atoms to convert some of the double bonds in unsaturated fatty acid residues into single bonds.

11.35 *Which two phospholipids are most prevalent in plants and animals?*

Phosphatidylcholine (lecithin) and phosphatidylethanolamine (cephalin) are the two phospholipids most prevalent in plants and animals.

11.37 *a. To which class of phospholipids does the compound belong?*

$$CH_3(CH_2)_{16}\overset{\overset{\displaystyle O}{\|}}{C}-O-CH_2$$

$$CH_3(CH_2)_5CH=CH(CH_2)_7\overset{\overset{\displaystyle O}{\|}}{C}-O-CH$$

$$CH_2-O-\underset{\underset{\displaystyle O^-}{|}}{\overset{\overset{\displaystyle O}{\|}}{P}}-OCH_2-\underset{\underset{\displaystyle NH_3^+}{|}}{CH}-\overset{\overset{\displaystyle O}{\|}}{C}-O^-$$

The compound shown is a glycerophospholipid because it consists of a glycerol molecule containing 2 fatty acids and phosphate group with an alcohol residue. It belongs to the phosphatidylserine family of glycerophospholipids.

b. How do other members of this class of phospholipids differ from one another?

The other members of this class have different alcohol components bonded to the phosphate group.

c. Draw the products obtained when the compound is saponified into its component parts.

See below.

d. For each product from part c, state whether it is hydrophobic, hydrophilic, or amphipathic.

$$CH_3(CH_2)_{16}-\overset{\overset{\textstyle O}{\|}}{C}-O^- \quad + \quad CH_3(CH_2)_5CH=CH(CH_2)_7-\overset{\overset{\textstyle O}{\|}}{C}-O^- \quad + \quad \begin{matrix} CH_2-OH \\ | \\ CH-OH \\ | \\ CH_2-OH \end{matrix}$$

amphipathic   amphipathic   hydrophilic

$$+ \quad O^--\overset{\overset{\textstyle O}{\|}}{\underset{\underset{\textstyle O^-}{|}}{P}}-O^- \quad + \quad HO-CH_2-\underset{\underset{\textstyle NH_3^+}{|}}{CH}-\overset{\overset{\textstyle O}{\|}}{C}-O^-$$

hydrophilic   hydrophilic

11.39 *How do phosphatidylethanolamines differ from phosphatidylcholines?*

The two families differ by the alcohol group attached to the phosphate group in a glycerophospholipid. In phosphatidylethanolamines, the alcohol is derived from ethanolamine: $HOCH_2CH_2NH_3^+$. In phosphatidylcholine, it is derived from choline:

$$HO-CH_2-CH_2-\overset{\overset{\textstyle CH_3}{|}}{\underset{\underset{\textstyle CH_3}{|}}{\overset{+}{N}}}-CH_3$$

11.41 *Draw a sphingomyelin that contains oleic acid.*

$$CH_3(CH_2)_{11}CH_2CH=CH-CH-OH$$

$$CH_3-(CH_2)_4-CH_2 \quad\quad CH_2-(CH_2)_6-\overset{\overset{\textstyle O}{\|}}{C}-NH-CH$$

$$\underset{\overset{|}{\underset{\textstyle H \quad H}{C=C}}}{}$$

$$CH_2-O-\overset{\overset{\textstyle O}{\|}}{\underset{\underset{\textstyle O^-}{|}}{P}}-OCH_2-CH_2-NH_3^+$$

11.43 *In certain membranes sphingomyelin serves as a replacement for phosphatidylcholine. What structural similarities between these classes of phospholipids make this a good substitution?*

Sphingomyelin and phosphatidylcholine have similar structures. They both contain a positively charged ammonium residue:

$$-CH_2-CH_2-\overset{+}{N}(CH_3)_3$$

$$-\overset{O}{\underset{O^-}{\overset{\|}{P}}}-O-CH_2-CH_2-\overset{+}{N}(CH_3)_3$$

phosphatidylcholine   sphingomyelin

11.45 *A particular cerebroside contains a glucopyransose residue attached to a sphingosine residue by a β glycosidic bond (see Section 10.6). Based on the sphingolipid shown in Figure 11.18, draw this cerebroside.*

$$CH_3(CH_2)_{11}CH_2CH=CHCH-OH$$

$$CH_3(CH_2)_5-CH=CH-(CH_2)_7-\overset{O}{\overset{\|}{C}}-NH-CH$$

(glucopyranose ring structure with CH$_2$OH, OH, HO, H substituents and O—CH$_2$ linkage)

11.47 *The sphingomyelins present in cow brain include residues of saturated fatty acids having between 16 and 32 carbon atoms and/or residues of unsaturated fatty acids with 18 or 22 carbon atoms. Draw one of these sphingomyelins.*

288

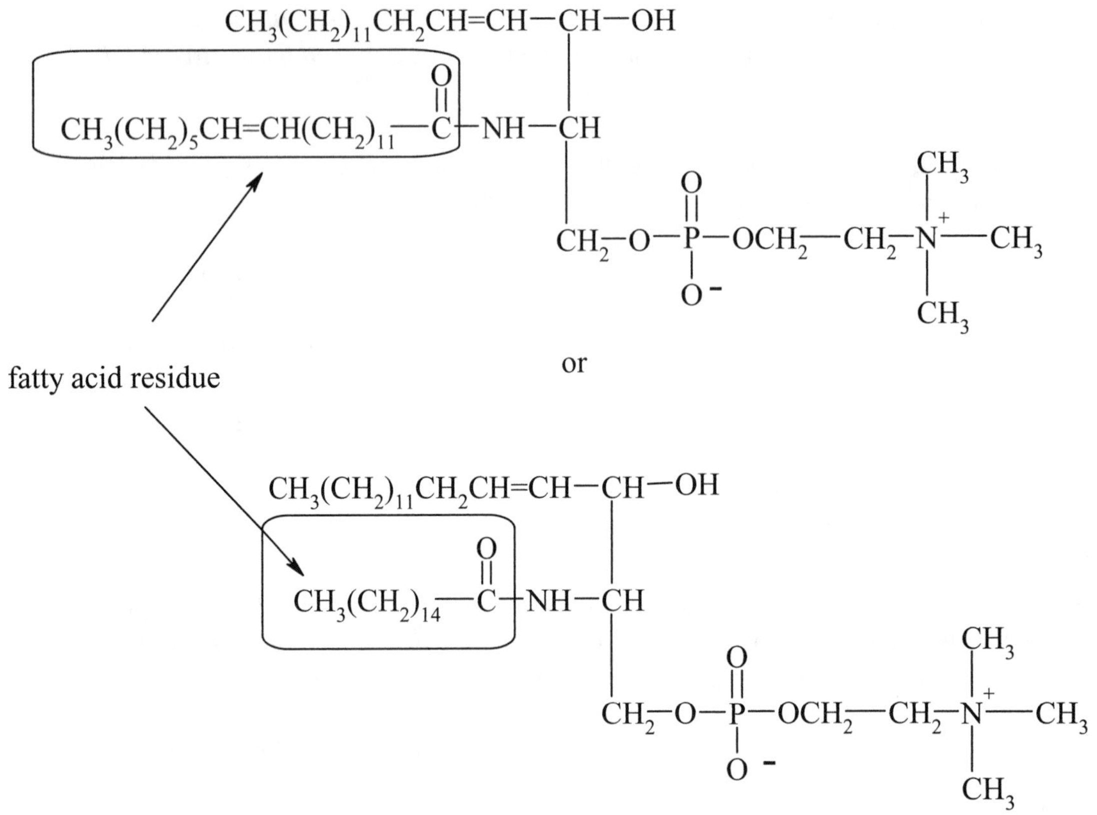

$$CH_3(CH_2)_{11}CH_2CH=CH-CH-OH$$

$$CH_3(CH_2)_5CH=CH(CH_2)_{11}-\overset{\overset{\displaystyle O}{\|}}{C}-NH-CH$$

$$CH_2-O-\overset{\overset{\displaystyle O}{\|}}{\underset{\underset{\displaystyle O^-}{|}}{P}}-OCH_2-CH_2-\overset{\overset{\displaystyle CH_3}{|}}{\underset{\underset{\displaystyle CH_3}{|}}{N^+}}-CH_3$$

fatty acid residue

or

$$CH_3(CH_2)_{11}CH_2CH=CH-CH-OH$$

$$CH_3(CH_2)_{14}-\overset{\overset{\displaystyle O}{\|}}{C}-NH-CH$$

$$CH_2-O-\overset{\overset{\displaystyle O}{\|}}{\underset{\underset{\displaystyle O^-}{|}}{P}}-OCH_2-CH_2-\overset{\overset{\displaystyle CH_3}{|}}{\underset{\underset{\displaystyle CH_3}{|}}{N^+}}-CH_3$$

11.49 *How is cholesterol used in the human body?*

The body uses cholesterol primarily as a precursor for the synthesis of other steroids.

11.51 *Label the hydrophobic and hydrophilic parts of taurocholate (Figure 11.21).*

Hydrophilic parts will have many –OH, –NH, or –O bonds while hydrophobic will be mostly long chains of hydrocarbons.

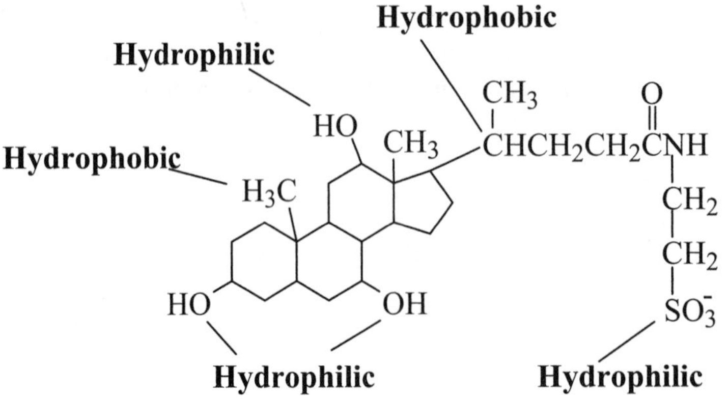

11.53 *Which molecule is the starting point for the synthesis of sex hormones and bile salts?*

Cholesterol.

11.55 *a. What are the biological functions of HDL and LDL?*

HDL transports cholesterol and phospholipids from cells back to the liver. LDL transports cholesterol and phospholipids from liver to cells.

b. In the news, you often hear HDL referred to as "good cholesterol" and LDL as "bad cholesterol". In terms of the structure HDL and LDL, is use of the term "cholesterol" totally correct? Explain.

Partially. Cholesterol is the major component of LDH. HDL contains more protein and phospholipids than cholesterol.

c. What makes HDL "good" and LDL "bad"?

Having high LDH and low HDL levels is a warning sign for atherosclerosis and an increased risk of stroke and heart disease.

11.57 *How do nonsteroidal anti-inflammatory drugs (NSAIDs) act to reduce pain, fever, and swelling?*

They block the action of COX-1 and COX-2, enzymes involved in the conversion of arachidonic acid into prostaglandins and thromboxanes.

11.59 *Cortisol, cortisone, and other anti-inflammatory steroids block the action of an enzyme that catalyzes the hydrolysis of unsaturated fatty acids, including arachidonic acid, from membrane lipids. How does this result in reduced inflammation?*

Hydrolysis of a particular phospholipid to release arachidonic acid is the first step in the production of prostaglandins. When cortisol, cortisone, and other anti-inflammatory steroids inhibit the enzyme that catalyzes this hydrolysis, the formation of prostaglandins is blocked.

11.61 *Dexamethasone, an alternative to COX-2 inhibitors, blocks the production of COX-2. What are the benefits of taking a drug that prevents formation of COX-2, but has no effect on COX-1?*

Drugs that prevent formation of COX-2 but have no effect on COX-1 avoid the side effects associated with the inhibition of COX-1 enzymes, namely ulcers and kidney damage.

11.63   *How are facilitated diffusion and active transport different?*

Facilitated diffusion moves substances from areas of higher concentration to areas of lower concentration, and does not require the input of energy. Active transport moves substances in the opposite direction and requires the input of energy.

11.65   *How are facilitated diffusion and diffusion different?*

In facilitated diffusion, diffusion across a membrane takes place with the assistance of proteins.

11.67   *To function properly, membranes must be flexible or fluid. In light of this fact, propose an explanation of why the cell membranes in the feet and legs of a reindeer contain a higher percentage of unsaturated fatty acids than do the cell membranes in the interior of its body.*

Reindeer are typically found in cold, snow-covered regions. Since the melting point of fatty acids goes down as the degree of unsaturation goes up, having more unsaturated fatty acids in the feet and legs would reduce the likelihood that the fats would solidify. This would help keep the membranes more flexible.

11.69   *Arachidonic acid (Figure 11.22) is not an omega-3 fatty acid. Which type of omega fatty acid is it?*

Omega-6 fatty acid. The first C=C bond from the free end of the hydrocarbon chain occurs at carbon-6.

11.71   *What are the health benefits of consuming omega-3 fatty acids?*

The known health benefits of consuming omega-3 fatty acids include: increased HDL; lowered LDL and triglycerides; decreased blood pressure; lowered risk for heart disease; and less rheumatoid arthritis inflammation.

11.73 *When unsaturated fatty acids were discussed in Section 11.1, it was said that their carbon-carbon double bonds are usually cis. Name one food source that is a natural source of unsaturated fatty acids with trans double bonds.*

Beef and dairy products contain natural *trans* fatty acids.

11.75 *Draw a* trans *fat that might form if the following triglyceride is subjected to partial hydrogenation*

$$CH_3(CH_2)_{12}\overset{\displaystyle O}{\overset{\|}{C}}-O-CH_2$$

$$CH_3(CH_2)_4CH_2\diagdown_{C=C}\diagup^{CH_2(CH_2)_6-\overset{\displaystyle O}{\overset{\|}{C}}-O-CH}_{H}$$

$$CH_3(CH_2)_4CH_2\diagdown_{C=C}\diagup^{CH_2(CH_2)_6-\overset{\displaystyle O}{\overset{\|}{C}}-O-CH_2}_{H}$$

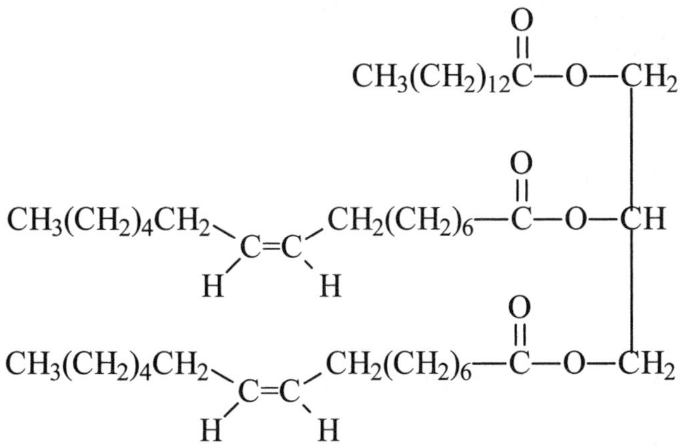

Hydrogenation is the chemical reaction that involves breaking the double bond between two carbons and adding a hydrogen atom to each carbon.

Since the question specifies partial hydrogenation, only one of the double bonds would be hydrogenated. The other bond undergoes *cis* to *trans* conversion.
 Two options are possible for the triglyceride in the problem. One of them is shown below.

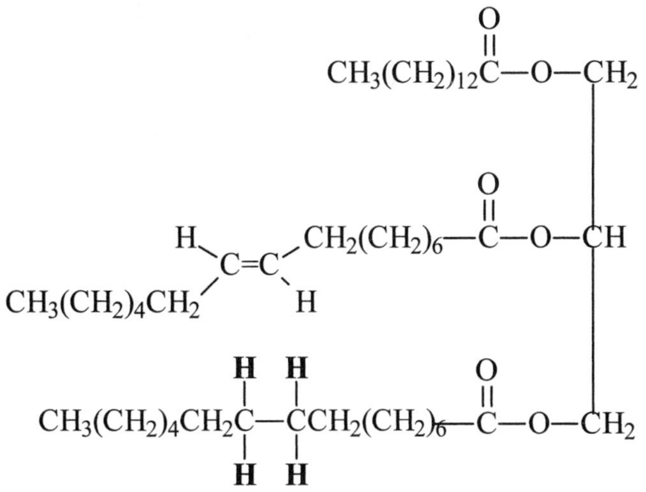

$$CH_3(CH_2)_{12}\overset{\displaystyle O}{\overset{\|}{C}}-O-CH_2$$

$$\overset{H}{\diagdown}_{C=C}\diagup^{CH_2(CH_2)_6-\overset{\displaystyle O}{\overset{\|}{C}}-O-CH}_{H}$$
$$CH_3(CH_2)_4CH_2\diagup$$

$$CH_3(CH_2)_4CH_2\overset{H\ \ H}{\underset{H\ \ H}{\overset{|\ \ |}{\underset{|\ \ |}{C-C}}}}CH_2(CH_2)_6-\overset{\displaystyle O}{\overset{\|}{C}}-O-CH_2$$

And the other is shown here.

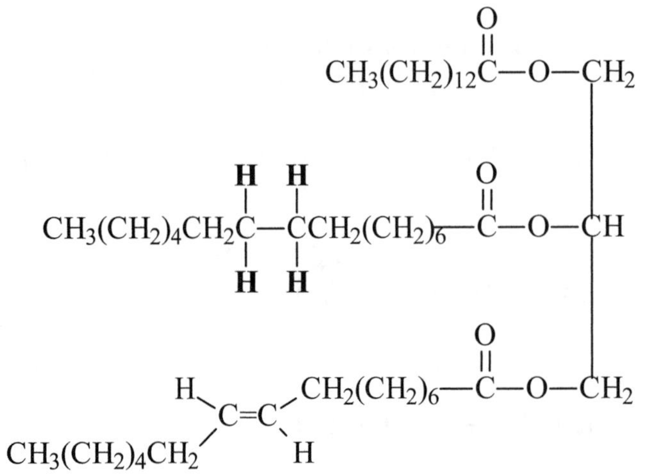

$$CH_3(CH_2)_{12}\overset{\overset{\displaystyle O}{\|}}{C}-O-CH_2$$

$$CH_3(CH_2)_4CH_2\overset{\overset{\textbf{H}}{\underset{\textbf{H}}{|}}}{C}-\overset{\overset{\textbf{H}}{\underset{\textbf{H}}{|}}}{C}CH_2(CH_2)_6-\overset{\overset{\displaystyle O}{\|}}{C}-O-CH$$

$$\underset{CH_3(CH_2)_4CH_2}{\overset{H}{\diagdown}}C=C\overset{CH_2(CH_2)_6-\overset{\overset{\displaystyle O}{\|}}{C}-O-CH_2}{\diagup}_{H}$$

11.77 *Draw the products formed if olestra (Figure 11.15) is saponified.*

During saponification, all ester bonds are hydrolyzed into carboxylate ions and alcohols.

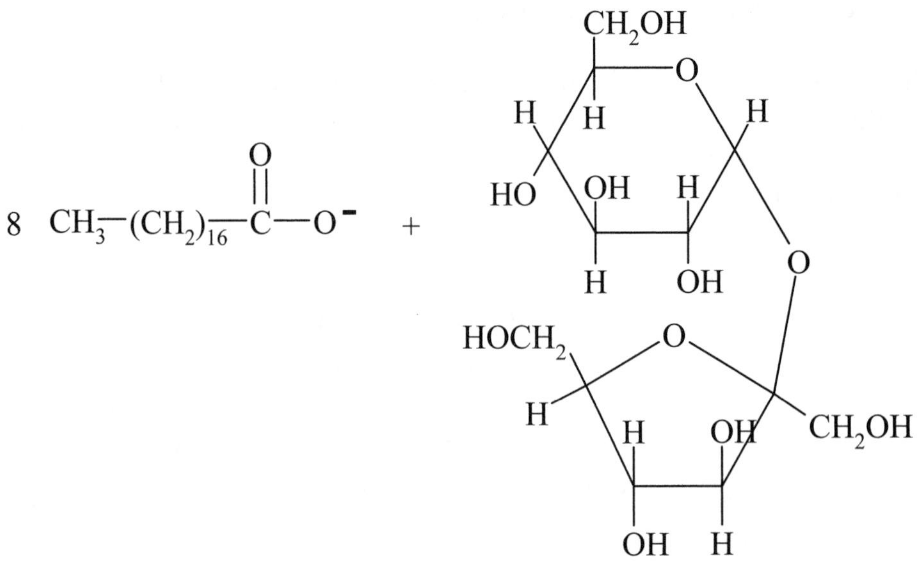

$$8 \quad CH_3\!-\!(CH_2)_{16}\!-\!\overset{\overset{\displaystyle O}{\|}}{C}-O^- \quad +$$

293

11.79 *Draw the hydrolysis products obtained when nandrolone laurate (Figure 11.29) is saponified.*

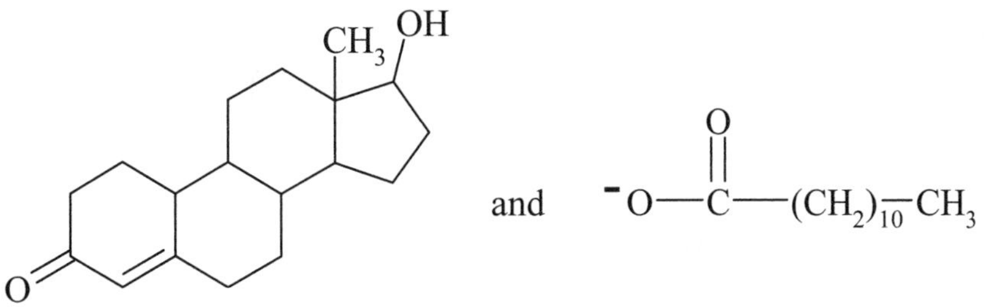

and $^-O-\overset{O}{\overset{||}{C}}-(CH_2)_{\overline{10}}CH_3$

11.81 *a. To which class of lipids does each molecule belong?*

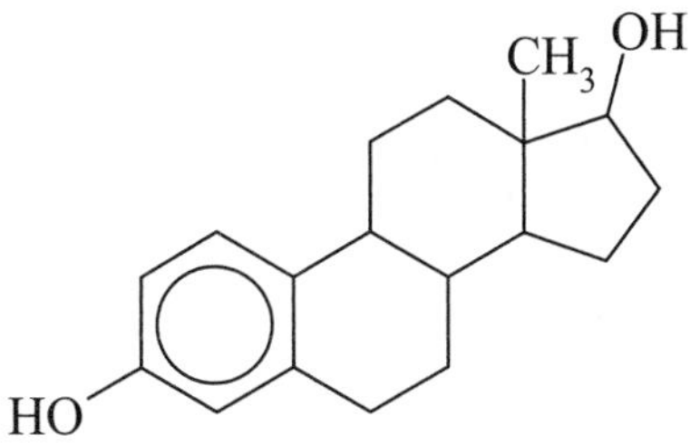

steroid

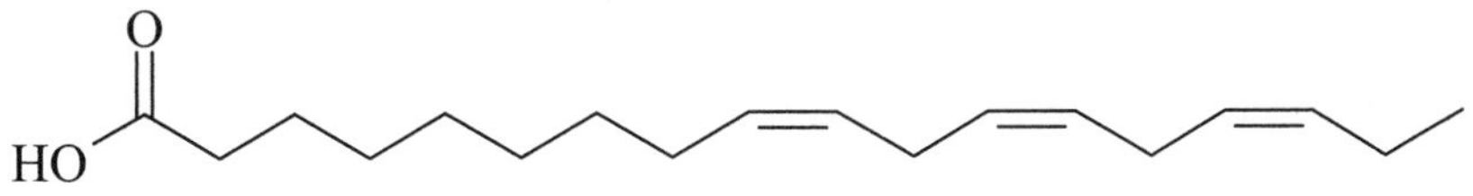

fatty acid

b. The top molecule in part a is estradiol. What is the function of this lipid?

Estradiol is a female sex hormone. Female sex hormones regulate menstruation, breast development, and other female traits.

c. Draw the product formed when the bottom molecule in part a is completely hydrogenated (reacted with 3H₂ and Pt).

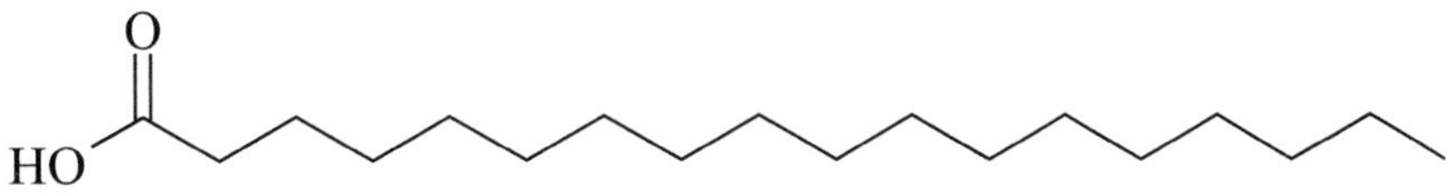

d. Which has a higher melting point, the bottom molecule in part a, or the molecule in your answer to part c?

The molecule in part *c* is expected to have a higher melting point because it is a saturated or a completely hydrogenated fatty acid.

e. In the body, some estradiol is combined with fatty acids to produce esters. It is believed that these esters serve as a way to store estradiol for later use. Using the two molecules in part a, draw one of these esters. Estradiol's alcohol group is involved in ester formation, not its phenol group.

f. Which would you expect is more hydrophilic, estradiol or the estradiol-containing ester in your answer to part e?

Estradiol. Estradiol has a smaller nonpolar or hydrophobic component. The long hydrocarbon chain in the estradiol-containing ester in part *e.* makes it more hydrophobic.

g. Estradiol fatty acid esters are carried through the blood in lipoproteins and are stored in body fat. Why are these esters soluble in fat?

Estradiol fatty acid esters have a significant hydrophobic (nonpolar) component.

11.83 *What is the chemical difference between vegetable oil and motor oil?*

Vegetable oil consists of triglycerides mostly made from unsaturated fatty acids. Motor oil is a mixture of different size long-chain hydrocarbons.

11.85 *Which fatty acid is predominantly involved in the biosynthesis of eicosanoids, compounds often involved in inflammatory processes?*

Arachidonic acid is the fatty acid used in the biosynthesis of eicosanoids.

Chapter 12
Peptides, Proteins, and Enzymes

Solutions to Problems

12.1   *Three amino acids are combined to produce a tripeptide.*

a. How many peptide bonds does the tripeptide have?

Two, as enclosed below.

b. Which amino acid residue is at the N-terminus?

c. Which amino acid residue is at the C-terminus?

d. How many different tripeptides could be produced using one of each amino acid?

6

e. How many different tripeptides could be produced using one, two, or three of each amino acid?

27

12.3   *Draw methionine as it would appear at each of the following pHs.*

a. pH 1

At a pH 1 the carboxyl group appears as $-CO_2H$ and the amino group as NH_3^+.

$$\overset{+}{H_3N}CH\overset{O}{\overset{\|}{C}}-OH$$
$$|$$
$$CH_2CH_2SCH_3$$

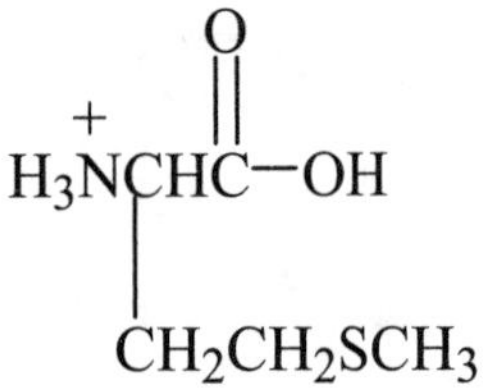

b. pH 7

At a pH 7 the carboxyl group appears as $-CO_2^-$ and the amino group as NH_3^+.

$$\overset{+}{H_3N}CH\overset{O}{\overset{\|}{C}}-O^-$$
$$|$$
$$CH_2CH_2SCH_3$$

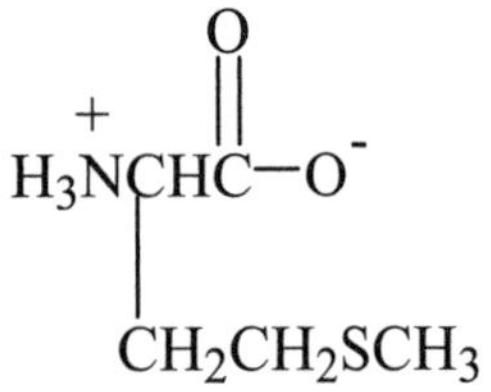

c. pH 14

At a pH 14 the carboxyl group appears as $-CO_2^-$ and the amino group as NH_2.

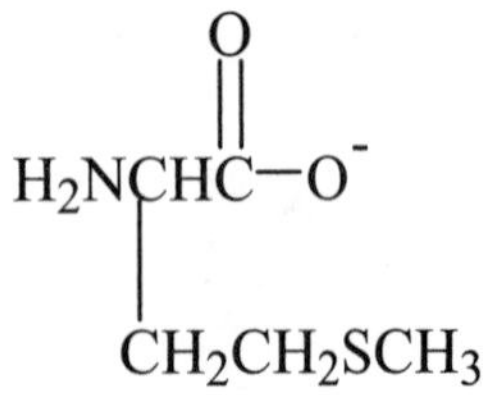

$$H_2NCH\overset{O}{\overset{\|}{C}}-O^-$$
$$|$$
$$CH_2CH_2SCH_3$$

12.5   *a. What is the net charge on arginine at pH 1?*

2+

At pH 1, the carboxyl group appears as $-CO_2H$ and the amino ($-NH_2$) group as NH_3^+. The side chain is basic and is positively charged.

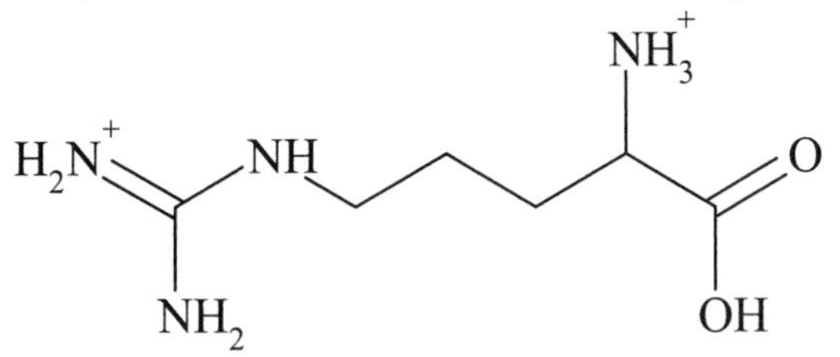

b. What is the net charge on arginine at pH 7?

1+

At pH 7, all the carboxyl groups are negatively charged and the amine groups are positively charged.

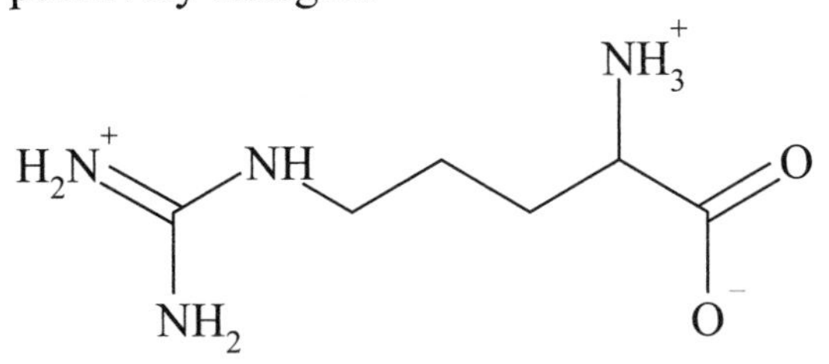

c. What is the net charge on arginine at pH 14?

1-

At pH 14, the carboxyl group appears as $-CO_2^-$ and the amino group as $-NH_2$. The side chain has lost a H^+ and has no charge.

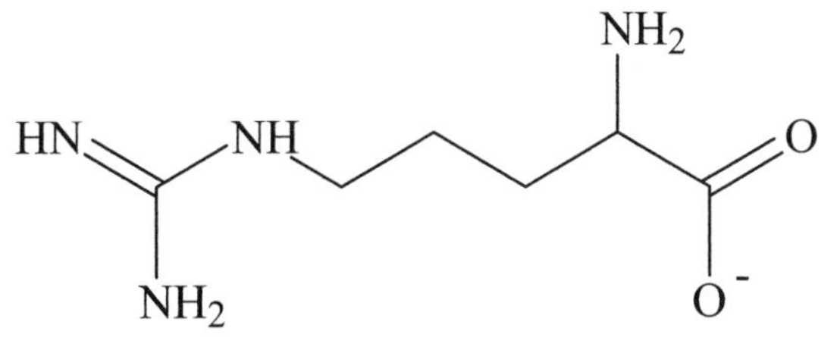

12.7   *Using Fischer projections draw each amino acid as it would appear at pH 7.*

A Fischer projection is drawn with the chiral carbon atom at the intersection of a vertical and a horizontal line. For amino acids, the carboxyl group points up, the

R group points down and the amino group points either left (L amino acid) or right (D amino acid). See Table 12.1 for the structure of each amino acid at pH 7.

a. L-isoleucine

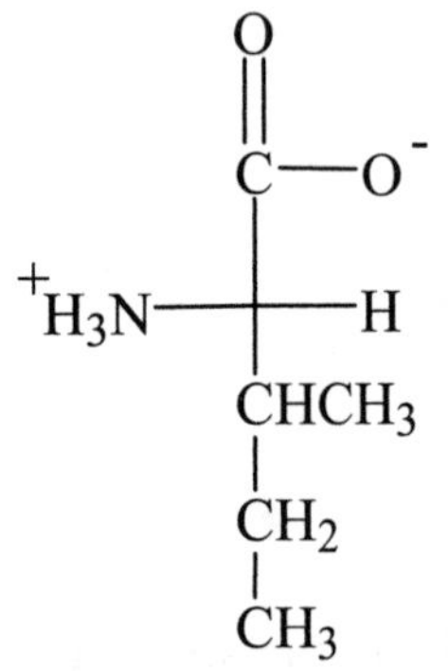

$$
\begin{array}{c}
O \\
\| \\
C-O^- \\
\overset{+}{H_3N}-\!\!\!-\!\!\!-H \\
CHCH_3 \\
| \\
CH_2 \\
| \\
CH_3
\end{array}
$$

b. D-aspartic acid

$$
\begin{array}{c}
O \\
\| \\
C-O^- \\
H-\!\!\!-\!\!\!-\overset{+}{N}H_3 \\
CH_2 \\
| \\
C-O^- \\
\| \\
O
\end{array}
$$

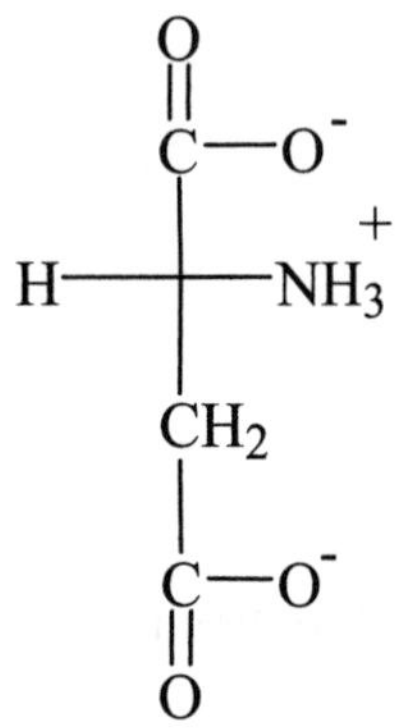

c. L-tyrosine

$$
\begin{array}{c}
O \\
\| \\
C-O^- \\
\overset{+}{H_3N}-\!\!\!-\!\!\!-H \\
CH_2 \\
\end{array}
$$

benzene ring with OH

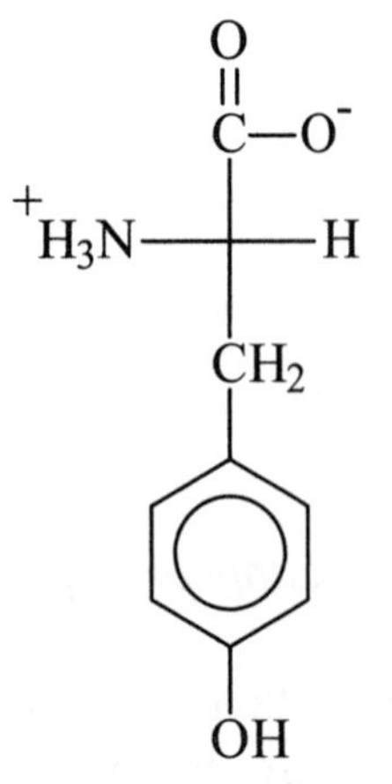

d. D-phenylalanine

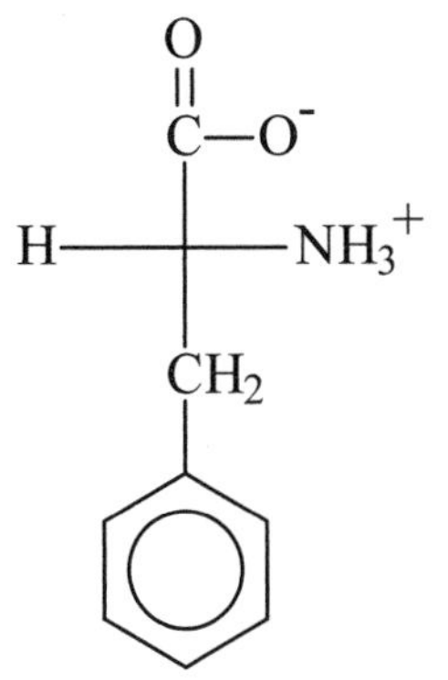

12.9 *Using a Fischer projection, draw each amino acid from Problem 12.7 as it appears at pH 1.*

At a pH of 1, carboxyl groups appear as $-CO_2H$ and amino groups as $-NH_3^+$.

a. L-isoleucine

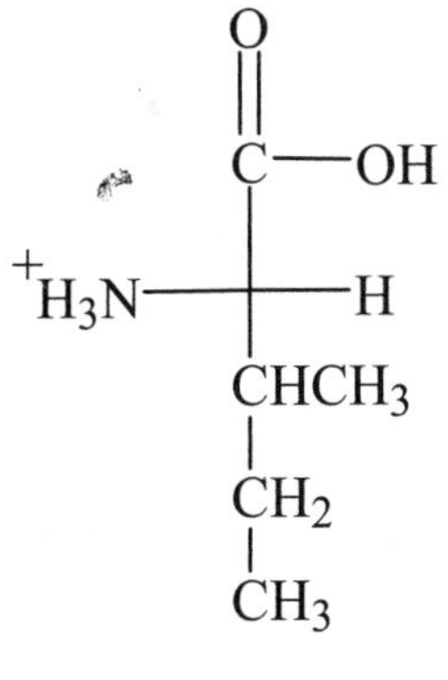

b. D-aspartic acid

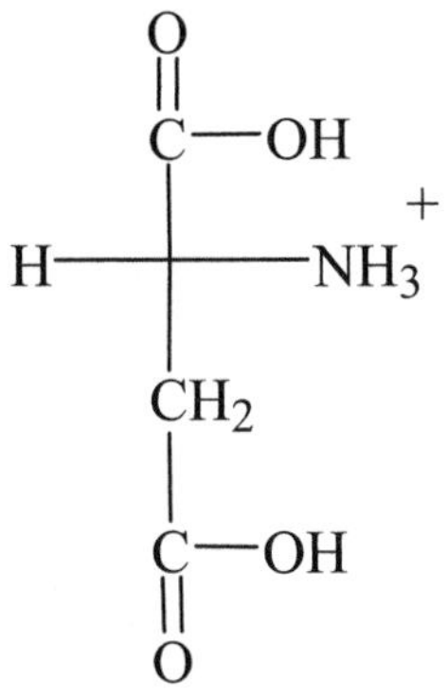

c. L-tyrosine

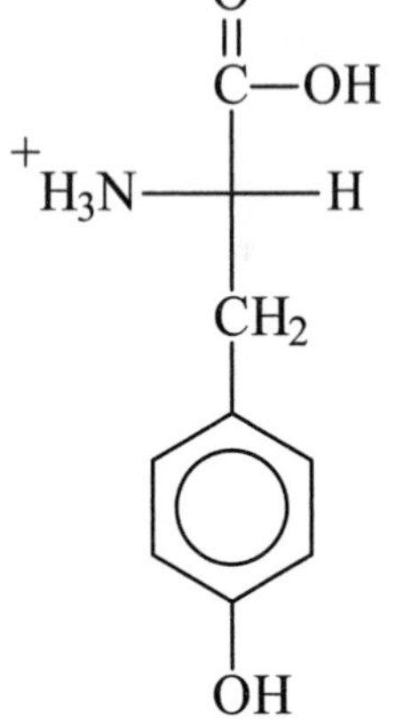

$$\begin{array}{c} O \\ \| \\ C-OH \\ | \\ ^+H_3N\!-\!\!\!-\!\!\!-\!\!H \\ | \\ CH_2 \\ | \end{array}$$

(benzene ring with OH)

d. D-phenylalanine

$$\begin{array}{c} O \\ \| \\ C-OH \\ | \\ H\!-\!\!\!-\!\!\!-\!NH_3^+ \\ | \\ CH_2 \\ | \end{array}$$

(benzene ring)

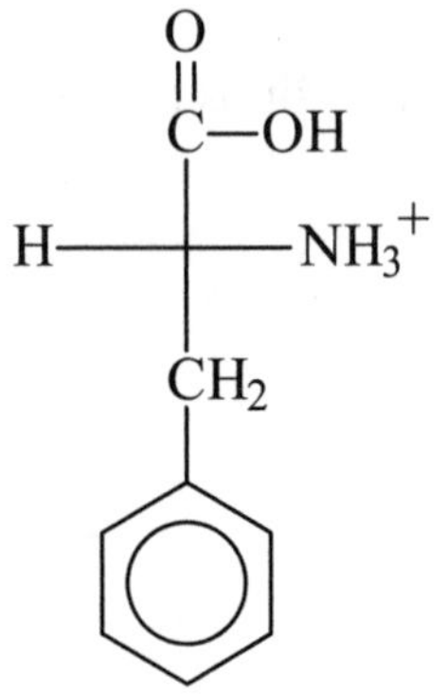

12.11 *Two of the amino acids in Table 12.1 have two chiral carbon atoms. Which ones?*

The two amino acids in Table 12.1 that have two carbons with four different atoms or groups of atoms, are threonine and isoleucine.

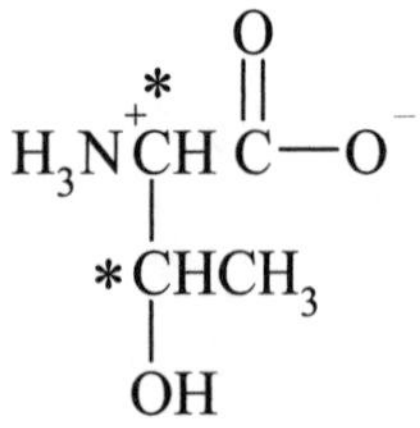

$$\begin{array}{c} * \quad O \\ \| \\ H_3N\overset{+}{C}H\ C\!-\!O^- \\ | \\ *CHCH_3 \\ | \\ OH \end{array}$$

threonine

$$\begin{array}{c} * \quad O \\ \| \\ H_3N\overset{+}{C}H\ C\!-\!O^- \\ | \\ *CHCH_3 \\ | \\ CH_2 \\ | \\ CH_3 \end{array}$$

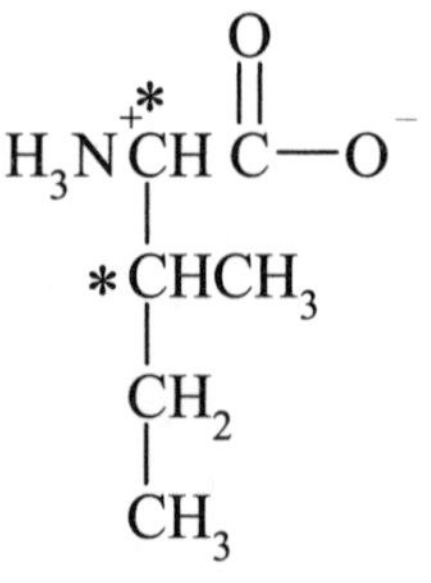

isoleucine

12.13 *Monosodium glutamate (MSG), used to enhance the flavor of certain foods, carries a net charge of zero. It can be formed by reacting glutamic acid as it appears at low pH (below) with sufficient NaOH. Draw MSG by showing the molecule below as it would appear at pH 7, and attaching Na^+ to the side chain.*

At a pH of 7, carboxyl groups appear as $-CO_2^-$ and the amino group as $-NH_3^+$.

$$
\begin{array}{c}
\overset{O}{\overset{\|}{}} \\
\overset{+}{H_3}NCH\overset{}{C}-O^- \\
|\\
CH_2 \\
|\\
CH_2 \\
|\\
C-O^- \ Na^+ \\
\|\\
O
\end{array}
$$

12.15 *Other amino acids than the twenty common ones shown in Table 12.1 appear in proteins. One of these, selenocysteine (Sec), is identical to cysteine except that the sulfur atom in Cys is replaced by a selenium atom.*

   a. *The side chain in Sec is as acidic as the side chain in Asp. Draw Sec as it would appear at pH 7.*

$$
\begin{array}{c}
\overset{O}{\overset{\|}{}} \\
H_3\overset{+}{N}CH\ C-O^- \\
|\\
CH_2Se^-
\end{array}
$$

   b. *To which classification of amino acids (nonpolar, polar-acidic, polar-basic, polar-neutral) does Sec belong?*

Polar-acidic, just like Asp.

12.17 *Which specific class of bonds holds one amino acid residue to the next in the primary structure of a protein? Are these bonds covalent or noncovalent?*

Peptide bonds, also known as amide bonds. Covalent.

12.19 *Draw the products obtained when Asp-Phe-OMe (Section 12.4) is hydrolyzed under acidic conditions.*

12.21 *a. Is Ala-Phe-Thr-Ser an oligopeptide or a polypeptide?*

Oligopeptide. The compound has four amino acid residues. Oligopeptides are defined to have between 2 and 10 amino acid residues.

b. How many peptide bonds does the molecule contain?

Three. A peptide bond is found between each of the joined residues.

c. Which is the N-terminal amino acid?

Alanine. The N-terminal residue is written first, therefore alanine is the N-terminal amino acid.

12.23 *Circle the backbone for the pentapeptide shown in Figure 12.4a.*

The backbone includes all of the peptide bonds.

12.25 *Name each of the amino acid residues in the oligopeptide shown in Figure 12.4a.*

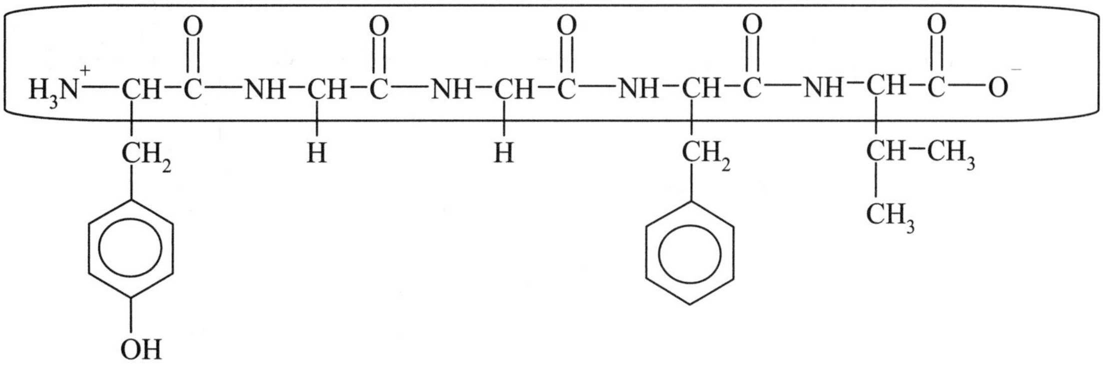

tyrosine   glycine   glycine   phenylalanine   valine

12.27. *Draw Gly-Phe-Lys-Lys as it would appear at*

a. pH 1

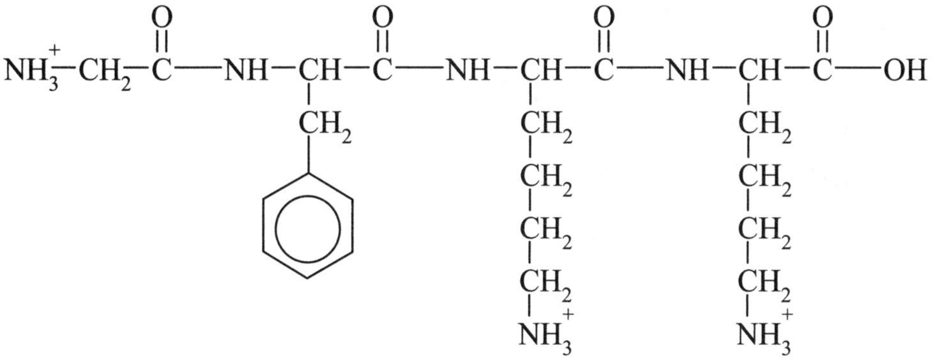

b. pH 7

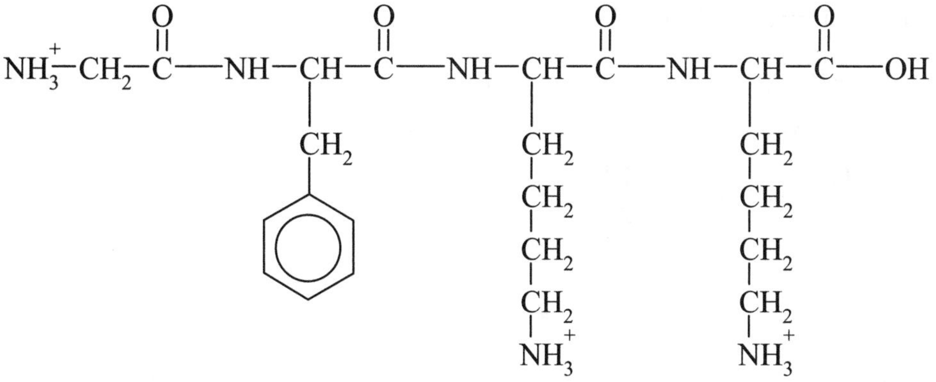

c. pH 14

$$NH_2-CH_2-C(=O)-NH-CH-C(=O)-NH-CH-C(=O)-NH-CH-C(=O)-O^-$$

with side chains:
- Phe: CH_2—(phenyl ring)
- Lys: $CH_2-CH_2-CH_2-CH_2-NH_2$
- Lys: $CH_2-CH_2-CH_2-CH_2-NH_2$

12.29 *What is the net charge on Asp-Lys at each pH?*

a. pH 1   **total charge = 2+**

$$\overset{+}{N}H_3-CH-C(=O)-NH-CH-C(=O)-OH$$

Asp side chain: $CH_2-C(=O)-OH$

Lys side chain: $CH_2-CH_2-CH_2-CH_2-\overset{+}{N}H_3$

b. pH 7   **total charge = 0**

$$\overset{+}{N}H_3-CH-C(=O)-NH-CH-C(=O)-O^-$$

Asp side chain: $CH_2-C(=O)-O^-$

Lys side chain: $CH_2-CH_2-CH_2-CH_2-\overset{+}{N}H_3$

c. pH 14   **total charge = 2-**

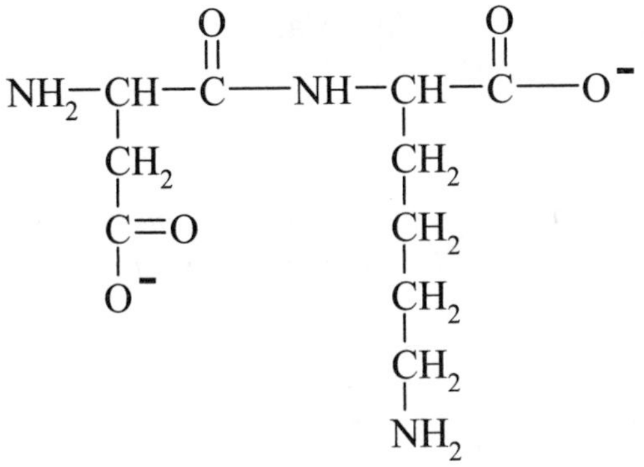

12.31   *What is the net charge on the oligopeptide in Figure 12.4a at*
a. pH 1
b. pH 14

The oligopeptide in Figure 12.4a has the following structure. Only the forms of the N-terminal amino group and the C-terminal carboxyl group are affected by pH.

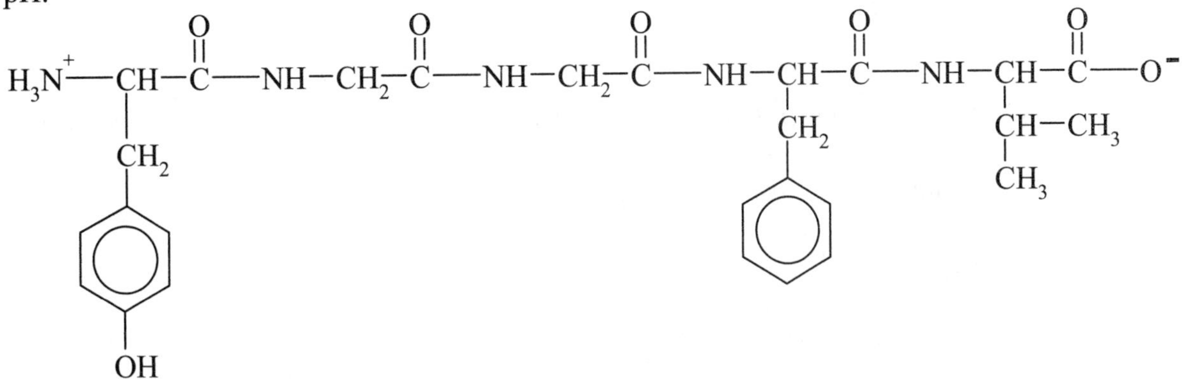

a. pH 1
1+
At pH 1, the carboxyl group at the C-terminal is in its acidic form (-CO_2H) and the amino group at the N-terminal is also in its acidic form (-NH_3^+). Therefore, there is a net charge of 1+.

b. pH 14
1-
At pH 14, the carboxyl group at the C-terminal is in its basic form (-CO_2^-) and the amino group at the N-terminal is also in its basic form (-NH_2). Therefore, there is a net charge of 1-.

12.33 *List the primary structure of all possible tripeptides containing one residue each of aspartic acid, phenylalanine, and valine.*

Asp-Phe-Val	Asp-Val-Phe	Phe-Asp-Val
Phe-Val-Asp	Val-Asp-Phe	Val-Phe-Asp

12.35 *In an aqueous environment will the following peptide fragment more likely be buried inside a globular protein or located on its surface? Explain.*

$$\text{~~NHCHC} \overset{O}{\underset{\|}{}} \text{—NHCHC} \overset{O}{\underset{\|}{}} \text{—NHCHC} \overset{O}{\underset{\|}{}} \text{—NHCHC} \overset{O}{\underset{\|}{}} \text{~~}$$

Side chains:
- CH_3CHCH_3
- CH_2 — (phenyl ring)
- $CHCH_3$ / CH_2 / CH_3
- CH_2 / CH_2 / S / CH_3

Inside. This arrangement allows the nonpolar side chains to interact through London forces and does not disrupt hydrogen bonding between water molecules.

12.37 *List the chemical bonds or forces that are primarily responsible for maintaining*

a. the primary structure of a protein

Peptide bonds

b. the secondary structure of a protein

Hydrogen bonds

c. the tertiary structure of a protein

Hydrogen bonds, salt bridges, hydrophobic effects, and disulfide bonds

d. the quaternary structure of a protein

Hydrogen bonds, hydrophobic effects, disulfide bonds, and salt bridges

12.39 *Which amino acids have side chains that can participate in salt bridge formation (ionic bonds) at pH 7.*

Aspartic acid, glutamic acid, lysine, and arginine.
A salt bridge is formed when a positively charged side chain is attracted to a negatively charged side chain. The only amino acids that have side chains that will have a charge at a pH of 7 are: arginine ($=NH_2^+$), aspartic acid ($-CO_2^-$), glutamic acid (CO_2^-), and lysine ($-NH_3^+$). Table 12.1 also lists histidine as having a charged side chain at pH 7 ($=NH^+-$). However, at pH 7, this side chain has *mostly* lost its H^+ and is therefore not charged.

12.41 *What distinguishes a protein that has quaternary structure from one that does not?*

A protein that has a quaternary structure consists of more than one polypeptide chain whereas a protein that has no quaternary structure consists of only a single polypeptide chain.

12.43 *Draw structures that show*

a. hydrogen bonding between His and Glu side chains.

hydrogen bonding

b. hydrogen bonding between Thr and Ser side chains.

hydrogen bonding

12.45 *Which is more likely to lead to a change in the biological activity of a protein, the replacement of a leucine residue with a valine residue or the replacement of the same leucine residue with a lysine? Explain.*

Both leucine and valine are nonpolar residues while lysine is a polar-basic residue. Replacing leucine with a lysine residue would more likely lead to a change in the three-dimensional structure of the protein and its biological activity. This is because the two residues differ in the types of noncovalent interactions in which they participate.

12.47 *What term is used to describe an enzyme whose tertiary structure has been unfolded?*

Denatured.

12.49 *Which types of bonds or interactions are disrupted during the denaturation of a protein?*

The bonds and interactions that are disrupted when a protein is denatured are those that contribute to the secondary, tertiary, and quaternary structures of the protein. These include: hydrogen bonds, salt-bridges, disulfide bonds, and hydrophobic interactions.

12.51 *Is it possible for an enzyme to show both absolute specificity and relative specificity?*

No. Absolute specificity means an enzyme can modify only one specific substrate whereas relative specificity allows an enzyme to accept a wider range of substrates that belong to a particular group. The range of substrates cannot be both narrow and wide.

12.53 *The enzyme glucokinase catalyzes the transfer of a phosphate group from ATP to carbon atom #6 of D-glucose. D-glucose is the only substrate for the enzyme.*

$$\text{D-glucose} + \text{ATP} \xrightarrow{\text{hexokinase}} \text{D-glucose-6-phosphate} + \text{ADP}$$

Which of the following terms describe the specificity of glucokinase: absolute specificity, relative specificity, stereospecificity?

Both absolute specificity and stereospecificity apply because only D-glucose is the substrate for the enzyme.

12.55 *The enzyme trypsin, found in the small intestine, operates best at pH 8. Draw a graph for trypsin, similar to that found in Figure 12.21a.*

At pH 8, the reaction rate would be highest.

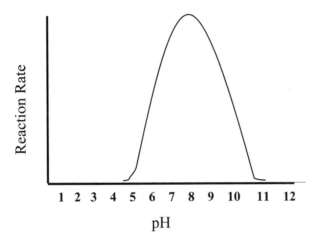

12.57 *A thermophilic bacterium found in a hot spring at Yellowstone Park produces an enzyme with a temperature optimum of 98°C. For this enzyme, draw a graph similar to that found in Figure 12.21b. Assume that the enzyme denatures at 100°C.*

The reaction rate peaks at the optimum temperature of 98 °C. At 100 °C, the enzyme denatures and loses its activity, thus a sharp drop-off of the reaction rate.

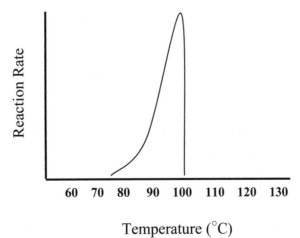

12.59 *The coenzyme NADH is a cofactor but is not a prosthetic group. Explain.*

NADH is a cofactor required by an enzyme for its catalytic action. It is not a prosthetic group because it is not bound to the enzyme protein and does not contribute to its tertiary structure.

12.61 *An inhibitor is added to a Michaelis-Menten enzyme. How could you distinguish between the inhibitor being reversible or irreversible?*

When an irreversible inhibitor acts on an enzyme, its effect on the catalytic activity of the enzyme is permanent. The activity of the enzyme can not be restored. When a reversible inhibitor binds to an enzyme, it interferes with the catalytic activity only on a temporary basis. When the inhibitor dissociates from the enzyme, the catalytic activity of the enzyme is restored.

12.63 *Explain why most competitive inhibitors are structurally similar to a substrate for the enzyme they inhibit, while most noncompetitive inhibitors are not.*

Competitive inhibitors bind at the substrate binding site. They can accomplish this easily if they are structurally similar to the substrate. Noncompetitive inhibitors do not bind at the substrate binding site.

12.65 *a. To which reaction type (synthesis, decomposition, single replacement, double replacement) does the reaction in Figure 12.24 belong?*

Double replacement. The acetyl group in aspirin switches places with the H atom in the –OH group of the serine side chain in COX.

b. Due to its action on COX enzymes, aspirin is considered to be a "suicide inhibitor". Explain.

In the process of transferring an acetyl group to COX, the aspirin itself is chemically altered and is no longer able to function as an inhibitor.

c. If transfer of an acetyl group from aspirin to the side chain of a serine residue in a COX enzyme caused a change in the tertiary structure of the enzyme, which interaction was most likely disrupted: a hydrogen bond, a salt bridge, or a disulfide bridge?

Hydrogen bond. The serine side chain lost its ability to hydrogen bond when the –OH group was replaced by the acetyl group.

12.67   *Celecoxib (Celebrex) is an inhibitor of COX-2 (Section 11.6). Given that arachidonic acid (Figure 11.22) is a substrate for COX-2, is celecoxib more likely to be a competitive inhibitor or a noncompetitive inhibitor?*

Celecoxib

Noncompetitive behavior. Because celecoxib and the arachidonic acid substrate do not have similar structures, their binding sites are probably different. Therefore, celecoxib is more likely a noncompetitive inhibitor.

12.69   *Describe the effect of positive and negative effectors on allosteric enzymes.*

A positive effector enhances substrate binding and increases the rate of the enzyme-catalyzed reaction. A negative effector reduces substrate binding and slows the reaction.

12.71   *a. Why are some enzymes produced initially as zymogens?*

Some enzymes are harmful to internal organs and may catalyze unwanted reactions. To control the activity of these enzymes, they are initially produced in their inactive form called zymogens. The zymogens are converted to their active, catalytic form when their specific substrates are present.

b. A person is found to produce trypsinogen molecules that have an abnormal primary structure. The trypsin produced from this zymogen is perfectly normal, both structurally and functionally. How is this possible?

The abnormal primary structure is on the part of the polypeptide chain removed during activation of the trypsinogen.

12.73 *List one unusual amino acid residue found in collagen. Is this residue present in a newly assembled collagen polypeptide? Explain.*

Hydroxyproline (Hyp). No, Hyp is formed by the enzyme-catalyzed oxidation of a hydroxyl group of a proline residue, but this occurs after the protein chain has been formed.

12.75 *Describe the quaternary structure of hemoglobin.*

A hemoglobin molecule is a tetramer consisting of four separate polypeptide chains (two identical α chains and two identical β chains). Each α chain consists of 141 amino-acid residues coiled into seven α-helical regions and each β chain consists of 146 amino acid residues coiled to form eight α-helical regions.

12.77 *True or false? If liver cells are the only cells in the human body that produce enzyme x, an increase in serum levels of enzyme x following a virus infection probably indicates that the infection has resulted in some liver damage. Explain.*

True. When cells die their contents are released into the bloodstream. If enzyme *x* is detected in the bloodstream in elevated concentrations it probably indicates liver cells have been damaged.

12.79 *Would orally administered asparaginase be as effective as intravenously injected asparaginase in reducing serum asparaginase levels? Explain.*

No. The asparaginase, like other proteins, would be broken down during digestion and would not enter the bloodstream.

12.81 *a. Give the complete name of each amino acid: Cys, Pro, Phe, and Val.*

cysteine, proline, phenylalanine, valine

b. Draw each of the amino acids in part a, as it would appear at pH 7.

c. Draw each of the amino acids in part a, as it would appear at pH 1.

Structure 1 (Cysteine):
$$H_3N^+-CH-C(=O)-OH$$
with side chain CH_2-SH

Structure 2 (Proline):
$$H_2N^+-CH-C(=O)-OH$$
with ring side chain $CH_2-CH_2-CH_2$ (pyrrolidine ring)

Structure 3 (Phenylalanine):
$$H_3N^+-CH-C(=O)-OH$$
with side chain CH_2—phenyl ring

Structure 4 (Valine):
$$H_3N^+-CH-C(=O)-OH$$
with side chain $CH-CH_3$ and CH_3

d. Draw each of the amino acids in part a, as it would appear at pH 14.

Structure 1 (Cysteine):
$$H_2N-CH-C(=O)-O^-$$
with side chain CH_2-SH

Structure 2 (Proline):
$$HN-CH-C(=O)-O^-$$
with ring side chain $CH_2-CH_2-CH_2$ (pyrrolidine ring)

Structure 3 (Phenylalanine):
$$H_2N-CH-C(=O)-O^-$$
with side chain CH_2—phenyl ring

Structure 4 (Valine):
$$H_2N-CH-C(=O)-O^-$$
with side chain $CH-CH_3$ and CH_3

e. The pentapeptide Cys-Pro-Phe-Val-Cys is known to inhibit the growth of Hantavirus. Draw this pentapeptide as it would appear at pH 7.

$$H_3N^+-CH-C(=O)-N-CH-C(=O)-NH-CH-C(=O)-NH-CH-C(=O)-NH-CH-C(=O)-O^-$$

Side chains:
- Cys: CH_2-SH
- Pro: ring $CH_2-CH_2-CH_2$
- Phe: CH_2—phenyl ring
- Val: $CH-CH_3$, CH_3
- Cys: CH_2-SH

f. Which amino acid residue is at the N-terminus? At the C-terminus?
Cysteine is at the N-terminus and cysteine is also at the C-terminus.

g. In its active form, the pentapeptide is bent into a horseshoe shape and held in that position by a disulfide bond. Draw this structure.

314

h. Does the pentapeptide have a primary structure? Explain.

Yes. The sequence of amino acids (Cys-Pro-Phe-Val-Cys) gives the primary structure.

i. If your answer to part h was yes, what holds the primary structure together?

Peptide bonds.

j. Does the pentapeptide have a secondary structure? Explain.

No. There is no H-bonding that takes place between amide N-H and C=O groups along the polypeptide backbone.

k. If your answer to part j was yes, what holds the secondary structure together?

No secondary structure.

l. Does the pentapeptide have a tertiary structure? Explain.

Yes. The overall three-dimensional shape of the pentapeptide is horseshoe shape.

m. If your answer to part l was yes, what holds the secondary structure together?

Disulfide bond.

n. Does the pentapeptide have a quaternary structure? Explain.

No. There is only one polypeptide chain.

o. If your answer to part n was yes, what holds the quaternary structure together?

No quaternary structure.

12.83 *As we will see in the next chapter, some of the information carried by DNA specifies the primary structure of proteins. Can a change in the structure of DNA (a change in the information) result in the altered tertiary structure of a protein that it carries the code for?*

Yes. The tertiary structure of a protein is determined by its primary structure.

12.85 *What is the name of this pentapeptide? (Use the three-letter abbreviations)*

Gln-Asp-Gly-Pro-Trp

Chapter 13
Nucleic Acids

Solutions to Problems

13.1   *The picture shows RNA polymerase attached to a strand of DNA*

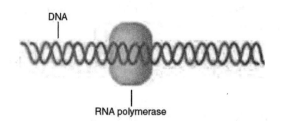

DNA

RNA polymerase

a. Which reaction is catalyzed by RNA polymerase?

Addition of nucleotide residues to a growing RNA strand.

b. What is the name of this process?

Transcription

c. Which building blocks are used to form the final product?

Ribonucleotides

d. What has to happen to the DNA double helix before RNA polymerase can catalyze this reaction?

The two strands of the DNA double helix have to separate.

e. What is the name of the DNA strand that gets "read" by RNA polymerase?

Template strand

f. In which direction does RNA polymerase move along the DNA?

3' to 5' on the template strand

g. To which end of the growing product chain are building blocks attached?

3' end

13.3 *a. Which monosaccharide is used to make DNA?*

2-Deoxyribose

b. Draw this monosaccharide in its β-furanose form.

The structure of β-2-deoxyribofuranose is given below:

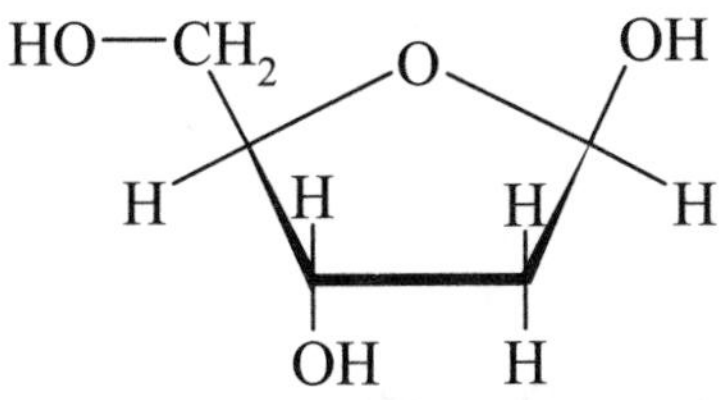

β-2-deoxyribofuranose

13.5 *Name the four bases that are present in DNA.*

Cytosine, guanine, thymine, and adenine

13.7 *Which of the bases present in nucleotides are purines?*

Adenine and guanine are purines.

13.9 *The structure of phosphoric acid is pH dependent.*

a. Draw the four forms in which phosphate appears.

The four forms of phosphate are:

phosphoric acid   dihydrogen phospate ion   hydrogen phosphate ion   phosphate ion

b. *Which form(s) appear at physiological pH?*

At physiological pH, phosphate exists in the forms of the dihydrogenphosphate ion and the hydrogenphosphate ion.

13.11 *Draw the phosphate monoester in Figure 13.1 as it would appear at low pH.*

$$CH_3O-\overset{\overset{\displaystyle O}{\|}}{\underset{\underset{\displaystyle OH}{|}}{P}}-OH$$

13.13 a. *How many different phosphate monoesters can β-D-ribofuranose form?*

Four phosphate esters can form because it has four -OH groups.

b. *When reacted with NH₂CH₃, how many different N-glycosides can β-D-ribofuranose form?*

One β-*N*-glycoside can form when an amine reacts at the hemiacetal carbon atom of the β-D-ribofuranose.

13.15 a. *What structural feature makes a base found in nucleic acids a purine?*

A purine is a base with the structure:

b. *Of the bases adenine, thymine, guanine, cytosine, and uracil, which are purines?*

Adenine and guanine are purines.

13.17   *Draw the complete structure of guanosine 5'-monophosphate (see Figure 13.6).*

The guanosine molecule is shown in Figure 13.5 and the nucleoside 5' -
monophosphate structure is shown in Figure 13.6. Draw the structure for
guanosine and then remove the –OH from carbon 5 of the ribose ring and replace
it with phosphate.

13.19   *How many phosphoester bonds and how many phosphoanhydride bonds are
present in the following?*

a. a nucleotide

One phosphoester bond and no phosphoanhydride bond.
A phosphoester bond connects a phosphate and a nucleoside. There is
one phosphate connected to the nucleoside and therefore one phosphoester bond.
There are no phosphoanhydride bonds (a chemical bond formed between two
phosphate groups) since there is only one phosphate (see Figure 13.6).

b. a nucleoside diphosphate

One phosphoester bond and one phosphoanhydride bond.
There is one phosphate connected to the nucleoside and therefore one
phosphoester bond. There are two phosphates and therefore one
phosphoanhydride bond (see Figure 13.7a).

13.21   *a. Draw 3'-dATP.*

3'-dATP is an abbreviation for 3'-deoxyadenosine triphosphate. Draw
deoxyadenosine. Remove the –OH on carbon 3 and add three phosphate groups.

b. Draw 3'ATP.

3'ATP is an abbreviation for adenosine triphosphate. Draw adenosine. Remove the –OH on carbon 3 and add three phosphate groups.

13.23   *a. Name the types of bonds broken when ATP is completely hydrolyzed.*

Phosphoanhydride, phosphoester, and N-glycoside (see Figure 13.7a).

b. What products are obtained when ATP is completely hydrolyzed?

3 phosphates, D-ribose, and adenine.

c. Name the types of bonds broken when cAMP is completely hydrolyzed.

Phosphoester and N-glycoside (see Figure 13.7b).

d. What products are obtained when cAMP is completely hydrolyzed?

1 phosphate, D-ribose, and adenine.

13.25   *a. Name the types of bonds broken when the molecule in Problem 13.21a is completely hydrolyzed.*

Phosphoanhydride, phosphoester, and N-glycosidic bonds are broken when the 3'-dATP molecule is completely hydrolyzed.

b. What products are obtained when this molecule is completely hydrolyzed?

Three phosphate ions, a β-D-2-deoxyribofuranose molecule, and an adenine molecule are obtained when the 3'-dATP molecule is completely hydrolyzed.

13.27   *In the sugar phosphate backbone of DNA*

a. which sugar residue is present?

2-Deoxyribose

b. which type of bonds connect the sugar and phosphate residues?

Phosphoester bonds

c. where are bases attached?

They are attached at C-1'.

13.29   *In terms of DNA and RNA structure, what do the terms 3'-terminus and 5'-terminus mean?*

At 3'- terminus of a DNA or RNA strand, the 3' hydroxyl group has no attached nucleotide residue. At the 5'-terminus, the 5' phosphate group has no attached nucleotide residue.

13.31 *Is each of the strands for the DNA in Figure 13.9 an oligonucleotide or polynucleotide?*

Oligonucleotide. Each DNA strand in Figure 12.9 consists of 10 nucleotide residues which makes each one an oligonucleotide.

13.33 *RNA consists of ribonucleotide residues connected to one another by what type of bond?*

Phosphodiester bonds.

13.35 *The term primary structure refers to the sequence of what in DNA?*

The term primary structure refers to the sequence of nucleotide residues in DNA.

13.37 *a. Draw the complete structure of the DNA dinucleotide dGC.*

The dinucleotide dGC consists of guanosine and cytidine groups bonded by a phosphoester bond between the 3' carbon of the guanosine and the 5' carbon of cytidine. See figure shown.

b. Label the 3' and 5' ends of the molecule.

The 5' end is the free phosphate group at carbon 5 of the guanosine residue and the 3' end is the free OH group at the 3' carbon of the cytidine residue.

13.39 *a. What does the term base pairing mean?*

The term base pairing refers to the hydrogen bonding that occurs between two complementary bases attached to the two sugar-phosphate backbones of a double-stranded DNA molecule.

b. In DNA, which bases are complementary to one another?

In DNA, cytosine (C) is complementary to guanine (G) and thymine (T) is complementary to adenine (A).

13.41 *Draw the dinucleotide that is complementary to dGC and label the 3' and 5' ends (see problem 13.37).*

The dinucleotide in Problem 13.37 has a guanosine residue at the 5' end, cytidine residue at the 3' end. For an antiparallel pairing, a dinucleotide complementary to this would also have a cytidine residue at the 3' end and a guanosine residue at the 5' end:

5' G ------ C 3' (Problem 13.37)
3' C ------ G 5' (complementary dinucleotide, see below)

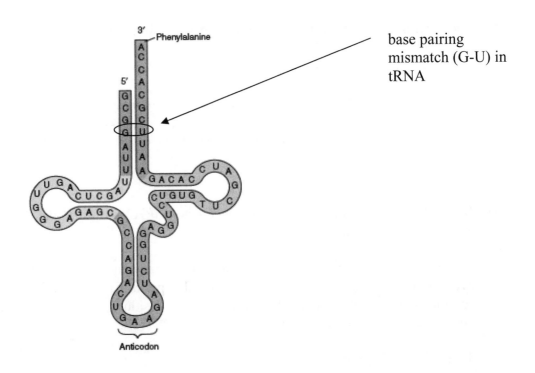

13.43 *Sometimes, not all of the bases in a region of double stranded DNA or RNA are paired with their complementary base. This often has little effect on the secondary and tertiary structure. Find an example of this base pairing mismatch in the tRNA molecule shown in Figure 13.17.*

13.45 *a. What are histones?*

Proteins that interact with the phosphate groups of the DNA backbone.

b. What are nucleosomes?

Groups of histones wrapped by DNA.

c. What is chromatin?

A coiled string of nucleosomes.

13.47 *When DNA is denatured, which of the following is disrupted?*
a. primary structure
b. secondary structure
c. tertiary structure

b. and *c.* only. When DNA is denatured, only the secondary (double-stranded helix held together by hydrogen bonds) and tertiary (overall three-dimensional shape including secondary structure) structures are disrupted.   The primary structure is not affected by denaturing.

13.49 *What molecule is made during the following processes?*

a. DNA replication

New DNA molecules

b. transcription

RNA molecules

c. translation

Protein molecules

d. reverse transcription

DNA molecules that contain genetic information for the synthesis of viral proteins

13.51 *Explain the term semiconservative replication.*

In semiconservative replication, a double-stranded DNA is synthesized from a daughter DNA strand and a parent DNA strand. The inclusion of a parent DNA as one of the two strands ensures the correct passing on of information contained in the parent DNA.

13.53 *What is a replication fork?*

A replication fork, found at each end of the origin, is where replication takes place.

13.55 *a. In which direction along an existing DNA strand does DNA polymerase move?*

Polymerases move along an existing DNA strand in the 3' to a 5' direction.

b. In which direction does DNA polymerase synthesize a new DNA strand?

Polymerases synthesize new DNA in the 5' to 3' direction.

c. If an origin opens and a DNA polymerase attaches to each of the exposed single strands of DNA, do the polymerases move in the same direction or in opposite directions as they make new DNA?

Opposite directions.

13.57 *Name the three types of RNA and describe their function.*

Transfer RNAs carry the correct amino acid to the site of protein synthesis.

Messenger RNAs carry the information that specifies which protein should be made.

Ribosomal RNAs combine with proteins to form ribosomes, the multi-subunit complexes in which protein synthesis takes place.

13.59 *a. In which direction does RNA polymerase move along a strand of DNA?*

From 3' to 5' on the DNA template strand

b. In which direction is RNA synthesized?

5' to 3'

13.61 *Enzymes called aminoacyl tRNA synthetases catalyze the combination of amino
acids with the proper tRNAs. The first step in this process is the reaction of an
amino acid with ATP to form an aminoacyl-AMP. An aminoacyl-AMP consists
of an amino acid residue attached through its carboxyl group to the 5' phosphate
group of AMP (Figure 13.6) by an anhydride bond. Draw the aminoacyl-AMP
that involves the amino acid alanine.*

13.63 *The sequence dGGCAT appears in a template strand of DNA.*

a. Which base is at the 3'-terminus of this primary structure?

Thymine.
Unless otherwise noted, a sequence of bases is listed from the 5'-terminus to the
3'-terminus. Therefore, the base at the 3'-terminus is thymine (T).

*b. What is the sequence in the DNA strand that is complementary to this
sequence? (Label the 3' and 5' ends.)*

 3'-terminus: dCCGTA : 5'-terminus
A and T are complementary bases, as are G and C. The complementary strand
will run from 3' to 5'.

*c. What is the sequence in the RNA strand that is synthesized from dGGCAT?
(Label the 3' and 5' ends.)*

3' terminus: CCGUA : 5' terminus

In the new RNA strand, U pairs with A in the DNA template and A pairs with T in the DNA template. G and C are complementary bases. The new RNA strand will run from 3' to 5'.

13.65 *a. What are codons?*

A codon is a series of three bases carried by mRNA that specifies a particular amino acid in the genetic code.

b. Which type of RNA has codons?

Messenger RNA

13.67 *Which amino acid is specified by each codon (listed 5' to 3')?*

See Table 13.1. Find the beginning code letter in the left column, locate the last code letter in the far right column. This establishes the row. Find the middle code letter at the top of the table and follow that column down to the row in line with your other two letters and locate the amino acid.

a. CCU

Pro

b. AGU

Ser

c. GUU

Val

d. GAA

Glu

13.69 *Which codon(s) specify each amino acid?*

Find the amino acid on Table 13.1. in the middle of the chart. The column in which it is located establishes the middle code letter. The left and right code letters are found on the same row in the left and right columns, respectively. Every row that contains the amino acid gives a different codon for it.

a. Phe

UUU and UUC

b. Lys

AAA and AAG

c. Asp

GAU and GAC

13.71   *Which anticodons are complementary to the codons in Problem 13.67?*

   a.   GGA (listed 3' to 5')
   b.   UCA (listed 3' to 5')
   c.   CAA (listed 3' to 5')
   d.   CUU (listed 3' to 5')

13. 73   *Name the components that combine with one another during the initiation of transcription.*

An mRNA and a tRNA combine with ribosome subunits.

13.75   *A tripeptide has the primary structure Ser-Lys-Asp.*

a. What is a sequence of bases in the mRNA that would code for this tripeptide? (Ignore start and stop codons and label the 3' and 5' ends.)

Ser has the following codons: UCU, UCC, UCA, UCG, AGU, AGC
Lys has the following codons: AAA, AAG
Asp has the following codons: GAU, GAC
A total of 24 sequences of bases in mRNA would code for Ser-Lys-Asp. They are shown below in the left column.

b. Assuming that no post-transcriptional modification took place, what is the sequence of bases in the template DNA strand used to make the mRNA? (Label the 3' and 5' ends.)

The corresponding sequence of template DNA strand is shown on the right column below. Any one of them would satisfy the requirement of writing a sequence.

330

(a) Possible sequences of mRNA bases			(b) Possible sequences of DNA bases	
(5' end) Ser – Lys – Asp	(3' end)		(3' end)	(5' end)
UCUAAAGAU			AGATTTCTA	
UCCAAAGAU			AGGTTTCTA	
UCAAAAGAU			AGTTTTCTA	
UCGAAAGAU			AGCTTTCTA	
AGUAAAGAU			UCATTTCTA	
AGCAAAGAU			UCGTTTCTA	
UCUAAGGAU			AGATTCCTA	
UCCAAGGAU			AGGTTCCTA	
UCAAAGGAU			AGTTTCCTA	
UCGAAGGAU			AGCTTCCTA	
AGUAAGGAU			TCATTCCTA	
AGCAAGGAU			TCGTTCCTA	
UCUAAAGAC			AGUTTTCTG	
UCCAAAGAC			AGGTTTCTG	
UCAAAAGAC			AGTTTTCTG	
UCGAAAGAC			AGCTTTCTG	
AGUAAAGAC			TCATTTCTG	
AGCAAAGAC			TCGTTTCTG	
UCUAAGGAC			AGATTCCTG	
UCCAAGGAC			AGGTTCCTG	
UCAAAGGAC			AGTTTCCTG	
UCGAAGGAC			AGCTTCCTG	
AGUAAGGAC			TCATTCCTG	
AGCAAGGAC			TCGTTCCTG	

13.77 *What is an operon?*

An operon is a group of genes whose expression is controlled by one promoter site. The transcription of this group of genes is initiated at the promoter site.

13.79 *Explain how allolactose is able to influence transcription of the lacZ, lacY, and lacA genes.*

Allolactose is a modified form of lactose. Its formation is catalyzed by a small amount of β-galactosidase in the cells when lactose is present. Allolactose binds to the repressor protein that controls the transcription of the *lacZ*, *lacY*, and *lacA* genes. Upon binding, allolactose modifies the shape of the repressor protein so that it is no longer able to attach to and repress the operator sites. Once the repressor protein is released from the operator sites, RNA polymerase can bind and begin the transcription of the three *lac* genes.

13.81   *Define the term mutation.*

A mutation is any permanent modification of the primary structure of DNA.

13.83   *Which of the levels of protein structure (primary, secondary, tertiary, quaternary) can be affected by a mutation?*

Only the primary structure (sequence of amino acids) can be affected by a mutation.

13.85   *Describe the general procedure used to make recombinant DNA and include the role of restriction enzymes, plasmids, and ligases.*

Recombinant DNA is formed by combining DNA segments from two different sources. A common type of recombinant DNA synthesis is the insertion of human DNA segments into bacterial plasmids. In this technique, the DNA segment of interest is cleaved from one source using a restriction enzyme specific to the isolation of the segment of interest. The same restriction enzyme can be used to cleave the bacterial plasmid at the appropriate site. Using ligases, these two segments attach to each other to form the recombinant DNA.

13.87   *What are short tandem repeats and how are they used in DNA fingerprinting?*

Short tandem repeats are relatively small stretches of DNA that contain short repeating sequences of bases. Restriction enzymes are used to cut the DNA into small segments and the number of base repeats of the STRs is determined.

13.89   *Beginning with one double strand of DNA, how many double-stranded DNAs will be present after 15 cycles of PCR?*

32,768 double-stranded DNAs
Each PCR cycle doubles the number of double-stranded DNAs. Therefore, after 15 cycles the number will be $2^{15} = 32,768$.

13.91 *Why do retroviruses have a high mutation rate?*

Retroviruses do not have proofreading and repair capabilities able to detect and fix changes in their base sequence. Therefore, they mutate rapidly.

13.93 *Describe the role of each in RNA interference.*

a. *double-stranded RNA*

RNA strands fold back on themselves to produce double-stranded RNA. This is the first step in RNA interference. The double-stranded DNA is the source of small interfering RNA (siRNA).

b. *nuclease*

Nuclease cuts double-stranded RNA into 21-23 base pair fragments called small interfering RNA.

c. *small interfering RNA*

Small interfering RNA's, the 21-23 base pair fragments created by nuclease, are the source of guide strands.

d. *guide strand*

A guide strand, one strand of each small interfering RNA, is complementary to the mRNA targeted for destruction.

e. *RISC*

RISC (RNA-Induced Silencing Complex) binds a guide strand to itself. When complementary mRNA binds to the guide strand, RISC cuts the mRNA, preventing it from undergoing translation and silencing the gene that produced the mRNA.

13.95 *a. Why do mutations in the BRCA1 or BRCA2 genes result in increased cancer risks?*

BRCA1 and BRCA2 genes code for proteins that are indirectly involved in recombinational repair of DNA. Mutations in these genes may stop production of these DNA repair proteins or modify the proteins into an inactive form. In either case, an increased risk of cancer results.

b. What is the normal function of the BRCA1 and BRCA2 genes?

The BRCA1 and BRCA2 genes code for proteins indirectly involved in a form DNA repair called recombinational repair.

13.97 *What is "sense" RNA and "antisense" RNA?*

A sense RNA is an RNA that would normally be translated to produce a particular protein. An antisense RNA is a complementary RNA that is transcribed from a new gene inserted into a piece of DNA. The antisense RNA base-pairs with the sense RNA locking it into a double-stranded RNA, preventing translation of the sense RNA and production of the protein.

13.99 *a. Draw and name the 5' deoxyribonucleotide formed by combining phosphate, β-2-deoxyribofuranose, and adenine.*

deoxyadenosine 5'-monophosphate

b. For this nucleotide, name the bond that joins the phosphate and deoxyribose residues and the bond that joins the deoxyribose and adenine residues.

See figure in part a.

c. Draw the dinucleotide formed by combining two of nucleotides from part a. Label the 5' and 3' end.

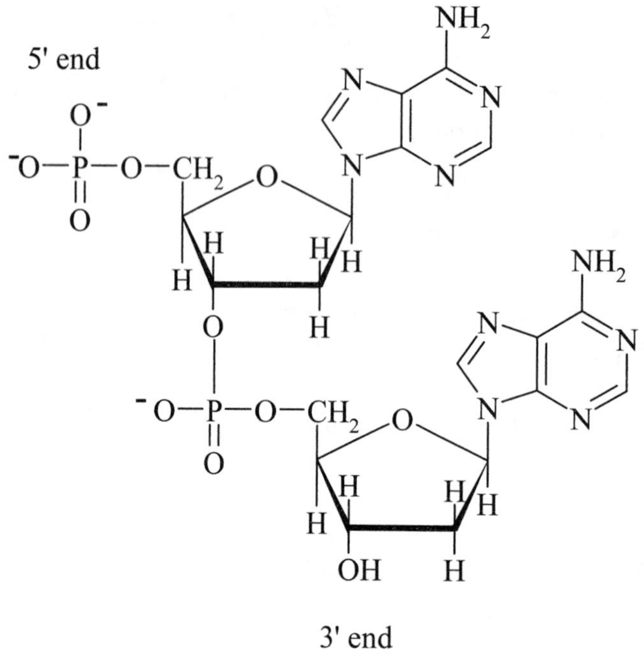

NH₂ → NH_2

5' end

O^-

$^-O-P-O-CH_2$

O

O

H

H H

H

O

H

NH_2

$^-O-P-O-CH_2$

O

O

H

H H

H

OH   H

3' end

d. For this dinucleotide, name the bond that joins the two nucleotide residues.

Phosphodiester bond.

e. Suppose that a trinucleotide is formed by adding a third nucleotide from part a to the dinucleotide from part c, and that this trinucleotide is incorporated into a DNA template strand that codes for mRNA. What codon will be transcribed from this trinucleotide residue?

UUU (listed 5'-3')

f. Which amino acid is specified by the codon in part e?

Phenylalanine

g. What is anticodon of the tRNA that carries the amino acid in part f?

AAA (listed 3'-5')

h. If a mutation changes base residue at the 5' end of the trinucleotide in part e from adenine to guanine, which new codon will be transcribed?
UUC (listed 5'-3')

i. Which amino acid is specified by the codon in part h?

Phenylalanine

j. What is the anticodon of the tRNA that carries the amino acid in part i?

(listed 3'-5') AAG

Chapter 14
Metabolism
Solutions to Problems

14.1   *In the drawing below, indicate where each of the following takes place.*

a. a 6 carbon molecule is converted into two 3 carbon compounds.

b. a 3 carbon compound loses one of its carbon atoms at the same time that it is activated.

c. a 4 carbon compound is converted into 6 carbon compound, which is then broken down into the same 4 carbon compound.

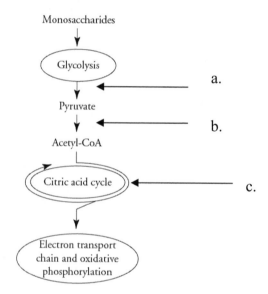

14.3   *Define the term metabolic pathway.*

A metabolic pathway is a group of biochemical reactions.

14.5   *In the first step of glycolysis, the following two reactions are coupled*

$$Glucose + P_i \rightarrow glucose\ 6\text{-}phophate + H_2O \qquad \Delta G = +3.3\ kcal$$

$$ATP + H_2O \rightarrow ADP + P_i \qquad \Delta G = -7.3\ kcal$$

a. Is the reaction Glucose + P_i → glucose 6-phosphate + H_2O spontaneous?

No. A positive ΔG ($\Delta G = +3.3$ kcal) indicates the reaction is nonspontaneous.

b. Write the net reaction equation and calculate ΔG for the coupled reaction.

The overall reaction is the sum of both reactions. Write ALL reacts on the left and ALL products on the right. Cancel out the reactants and products that are the same on both sides. Write the remaining reactants and products as the overall reaction. Add the ΔG values, $+3.3$ kcal $+ (-7.3$ kcal$) = -4.0$ kcal.

glucose + ~~H_2O~~ + ATP + ~~P_i~~ $\rightarrow$ glucose 6-phosphate + ADP + ~~P_i~~ + ~~H_2O~~

glucose + ATP $\rightarrow$ glucose 6-phosphate + ADP $\Delta G = -4.0$ kcal

c. Is the first step in glycolysis spontaneous?

Yes. A negative ΔG ($\Delta G = -4.0$ kcal) indicates the first step in glycolysis is spontaneous.

14.7   *Suppose that an uncatalyzed reaction is spontaneous because ΔG has a value of -10 kcal/mol. An enzyme that catalyzes the reaction is identified. What effect will the enzyme have on the rate of the reaction?*

A catalyst lowers the activation energy. A lower activation energy increases the rate of reaction. The catalyst does not lower the ΔG value, however.

14.9   *What is metabolism?*

Metabolism is the sum of all reactions that take place in a living thing.

14.11   *What provides the energy used to produce ATP from ADP and P_i, anabolism or catabolism?*

Catabolism

14.13   *Define the term reduction and explain how it applies to the difference in the structures of NAD^+ and NADH.*

Reduction is the gain of electrons. In organic and biochemical molecules, a gain of hydrogen and /or a loss of oxygen is indication that reduction has taken place. NADH, the reduced form of the coenzyme, has one more hydrogen atom than NAD^+, the oxidized form of the coenzyme.

14.15 *Describe the role of the oxidized and reduced forms of NAD⁺ and FAD in metabolism.*

In the oxidation reactions that take place in catabolism, the oxidized coenzymes NAD^+ and FAD are converted to their reduced forms, NADH and $FADH_2$, by accepting hydrogen atoms. The energy of these reduced forms can be used to produce ATP and the hydrogen atoms can be used during anabolism.

14.17 *The abbreviation P_i is used to represent phosphate.*

a. Draw phosphate ion.

The phosphate ion has the following structure:

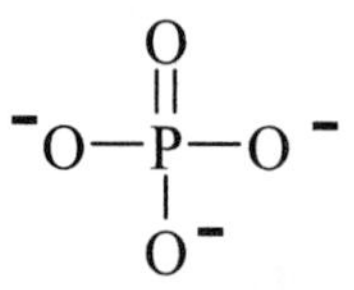

$$\begin{array}{c} O \\ \| \\ {}^-O-\overset{}{\underset{\underset{O^-}{|}}{P}}-O^- \end{array}$$

b. Depending on the pH of a solution, phosphate ion can exist in three other forms. Draw and name them.

The three other forms of phosphate ion vary according to the number of hydrogen atoms attached to the oxygen:

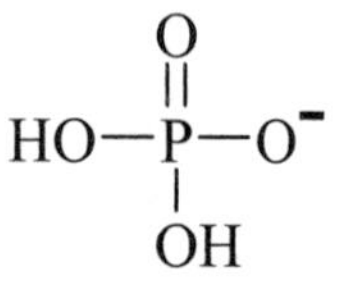

$$\begin{array}{ccc}
\begin{array}{c} O \\ \| \\ HO-\overset{}{\underset{\underset{OH}{|}}{P}}-O^- \end{array} &
\begin{array}{c} O \\ \| \\ HO-\overset{}{\underset{\underset{O^-}{|}}{P}}-O^- \end{array} &
\begin{array}{c} O \\ \| \\ HO-\overset{}{\underset{\underset{OH}{|}}{P}}-OH \end{array}
\end{array}$$

hydrogen phosphate ion 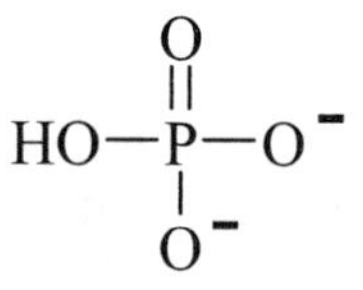   dihydrogen phosphate ion 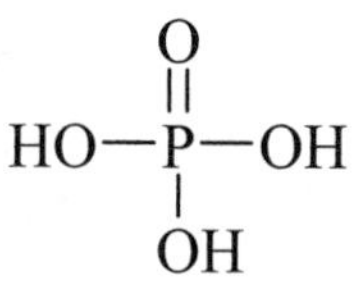   phosphoric acid

14.19 *Write a reaction equation for the hydrolysis of GTP to produce P_i.*

$GTP + H_2O \rightarrow GDP + P_i$

14.21 *What products are formed when each of the following undergoes digestion and is broken down to its building blocks?*

a. a polysaccharide

Monosaccharides are formed when a polysaccharide is broken down during digestion.

b. a triglyceride

Fatty acids, glycerol, and monoacylglycerides are produced when triglycerides undergo digestion.

c. a protein

Amino acids are produced when a protein is broken down to its building blocks.

14.23 *Name the functional group hydrolyzed when each undergoes digestion.*

a. a polysaccharide

Glycosidic bonds are hydrolyzed.

b. a triglyceride

Ester functional groups are hydrolyzed.

c. a protein

Amide functional groups are hydrolyzed.

14.25 *a. How do the structures of amylose and amylopectin, the two homopolysaccharides that make up starch, differ?*

Amylopectin contains α-(1 → 6) glycosidic bonds. Amylose does not.

b. How are the structures of amylose and amylopectin, the two homopolysaccharides that make up starch, similar?

Both amylose and amylopectin consist of glucose units connected by α(1→4) glycosidic bonds.

14.27 *Name three common disaccharides and name the monosaccharide(s) formed when they are hydrolyzed in the small intestine.*

Upon hydrolysis, the following disaccharides will form monosaccharides as shown:

Sucrose → glucose + fructose
Maltose → 2 glucose
Lactose → glucose + galactose

14.29 *What is the role of bile acids in digestion?*

Bile acids are emulsifying agents in the digestion of triglycerides in the small intestine.

14.31 *Studies have shown that, in newborns, pancreatic lipases are poor catalysts. Lingual lipase, whose secretion is stimulated by feeding, is active, however. Explain how the digestion of triglycerides by newborns differs from that of adults.*

In newborns, the hydrolysis of triglycerides begins in the stomach using lingual lipase, whereas in adults the triglyceride digestion occurs mostly in the small intestine using pancreatic lipases.

14.33 *a. What is the role of glycolysis?*

Glycolysis converts a glucose molecule into 2 pyruvate ions with accompanying production of 2 ATP and 2 NADH.

b. What is the net change in ATp and NADh from the passage of one glucose molecule through glycolysis?

2 ATP molecules and 2 NADH molecules are produced from one glucose molecule that undergoes glycolysis.

14.35 *a. What is the relationship between glucose 6-phosphate and fructose 6-phosphate (step 2 of glycolysis)? Are they constitutional isomers, cis/trans isomers, stereoisomers, different conformations of the same compound, or identical compounds?*

Constitutional isomers. Glucose 6-phosphate and fructose 6-phosphate have the same chemical formulas but different atomic connections.

b. Are these two compounds reducing sugars? Explain.

Yes, the two compounds are reducing sugars. Glucose 6-phosphate opens up into an aldose form and fructose 6-phosphate opens up into a ketose form which is transformed into an aldose in the basic conditions of the Benedict's reagent

c. Draw each as it appears in its open form.

open form of glucose 6-phosphate   open form of fructose 6-phosphate

14.37 *a. In humans, pyruvate is converted into what compound under anaerobic conditions?*

Under anaerobic conditions, pyruvate is reduced to lactate in humans.

b. What purpose does this reaction have?

Reduction of pyruvate to lactate results in the oxidation of NADH under anaerobic conditions to produce the NAD^+ required to continue glycolysis. The lactate is sent to the liver and used to make glucose.

14.39 *What is the net change in ATP and NADH from the passage of one glucose molecule through glycolysis, followed by the conversion of pyruvates into lactates?*

2 ATP molecules and 2 NADH molecules are produced when one glucose molecule goes through glycolysis and produces 2 pyruvate molecules. The conversion of the 2 pyruvate ions to 2 lactate ions uses up the 2 NADH formed in glycolysis. The net change is a gain of 2 ATP and no change in NADH.

14.41 *Draw pyruvic acid, the acidic form of pyruvate. Which predominates at physiological pH, the acidic form or the conjugate base form of this acid?*

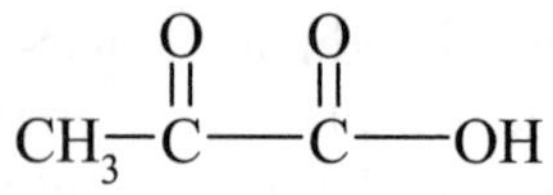

$$CH_3-\overset{\overset{\displaystyle O}{\|}}{C}-\overset{\overset{\displaystyle O}{\|}}{C}-OH$$

pyruvic acid

Physiological pH (7.4) is a slightly basic condition. Under this pH, the conjugate base form of pyruvic acid will predominate:

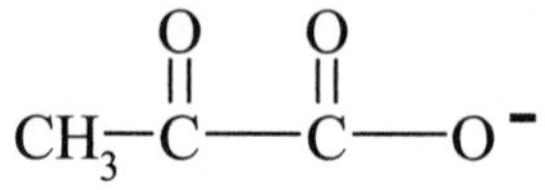

$$CH_3-\overset{\overset{\displaystyle O}{\|}}{C}-\overset{\overset{\displaystyle O}{\|}}{C}-O^-$$

pyruvate ion

14.43 *a. Is the conversion of acetaldehyde to ethanol an oxidation or a reduction?*

Reduction. The conversion of acetaldehyde to ethanol is shown below. This reaction is a reduction reaction because the carbonyl group is reduced to a single bond to oxygen with an increase in bonds to hydrogen atoms.

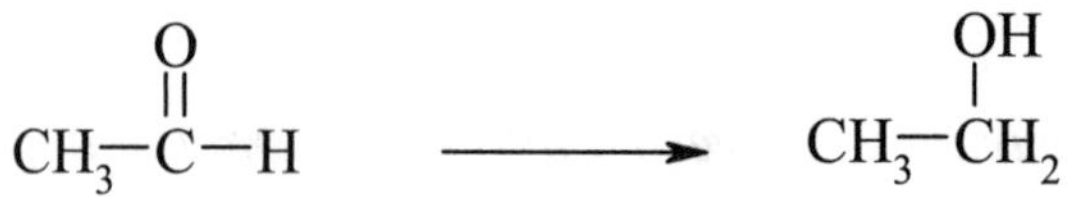

$$CH_3-\overset{\overset{\displaystyle O}{\|}}{C}-H \longrightarrow CH_3-\overset{\overset{\displaystyle OH}{|}}{CH_2}$$

acetaldehyde   ethanol

b. When this reaction takes place during alcoholic fermentation, what is the oxidizing agent?

acetaldehyde

c. What is the reducing agent?

NADH

14.45 *a. In what part of a cell does glycolysis take place?*

In the cytoplasm of the cell.

b. In what part of a cell does the conversion of pyruvate into acetyl-CoA take place?

In the mitochondria of the cell.

14.47   *Glycolysis can be described as a process in which energy is invested "up front" in exchange for a greater return of energy later on. Which steps in glycolysis involve investment of energy?*

Steps 1 and 3 involve "investment" or input of energy through coupling of the reaction with ATP hydrolysis.

14.49   *Monosaccharides other than glucose are converted into compounds that are intermediates in glycolysis. Why is this more efficient than having a different catabolic pathway for the conversion of each monosaccharide into pyruvate?*

Each different pathway would require a different set of enzymes.

14.51   *a. Define the term gluconeogenesis.*

Gluconeogenesis is a metabolic pathway that synthesizes glucose from noncarbohydrate sources such as amino acids, lactate, and glycerol.

b. What is the role of this pathway?

One role of this pathway is to recycle lactate into glucose. This pathway is the main source of glucose during fasting or starvation.

14.53   *Which three steps in glycolysis cannot be directly reversed during gluconeogenesis?*

Steps 1, 3, and 10.

14.55   *From which intermediate shared by glycolysis and gluconeogenesis is glycogen produced?*

Glucose 6-phosphate

14.57   *a. Describe the role that insulin plays in glycogenesis and glycogenolysis.*

Insulin activates glycogen synthetase (speeds up glycogenesis) and deactivates glycogen phosphorylase (slows down glycogenolysis).

b. Describe the role that glucagon plays in glycogenesis and glycogenolysis.

Glucagon deactivates glycogen synthetase (slows down glycogenesis) and activates glycogen phosphorylase (speeds up glycogenolysis).

14.59 *In the first step of the citric acid cycle, citrate (6 carbon atoms) is formed. The final product of the cycle is oxaloacetate (4 carbon atoms). Where do the missing two carbon atoms end up?*

The 2 carbon atoms are released as CO_2 molecules in steps 3 and 4.

14.61 *a. Which step is the control point of the citric acid cycle?*

Step 3 of the citric acid cycle, the conversion of isocitrate to α-ketoglutarate, acts as the control point for this cycle.

b. Name the enzyme that catalyzes the reaction that takes place at this step and list its positive and negative effectors.

The enzyme that catalyzes this reaction is isocitrate dehydrogenase. The negative effectors for this allosteric enzyme are ATP and NADH and the positive effectors are ADP and NAD^+.

14.63 *a. Which two compounds donate their electrons to electron transport chain?*

NADH and $FADH_2$

b. Which molecule is the final electron acceptor of the electron transport chain?

The O_2 molecule.

14.65 *As electrons pass through the electron transport chain, H^+ is moved to the mitochondrial intermembrane space. Which term best describes this process: diffusion, facilitated diffusion, or active transport?*

H^+ moves to the mitochondrial intermembrane space by active transport.

14.67   a. *The electron transport chain and oxidative phosphorylation typically generate how many ATP from 1 NADH?*

2.5 moles of ATP's are produced for every 1 mole of $NADH_2$ that goes through the electron transport chain and oxidative phosphorylation.

b. The electron transport chain and oxidative phosphorylation typically generate how many ATP from 1 FADH₂?

1.5 moles of ATP's are produced for every 1 mole of $FADH_2$ that goes through the electron transport chain and oxidative phosphorylation.

14.69   *Glycerol can be converted into a compound used in glycolysis and gluconeogenesis. Explain.*

Glycerol is produced by the hydrolysis of triglycerides. It can be converted into glycerol-3-phosphate which is transformed into dihydroxyacetone phosphate, an intermediate in both glycolysis and gluconeogenesis.

14.71   *In one pass through the β oxidation spiral a fatty acyl-CoA is shortened by two carbon atoms. What other products are formed?*

One acetyl-CoA, one $FADH_2$, and one NADH

14.73   *Calculate the net number of ATPs produced when one 14-carbon fatty acid is activated, undergoes β oxidation, and the products pass through the citric acid cycle and the electron transport chain.*

Activation in the first step requires the energy equivalent of 2 ATP's (see explanation in Section 15.9). Each step in the spiral produces one $FADH_2 = 1.5$ ATP and one NADH = 2.5 ATP. This represents ATP production with each pass through the cycle. The number of passes = (number of carbons – 2)/2 so for a 14-carbon fat there are (14-2)/2 = 6 passes with an acetyl-CoA produced on the final pass.

FADH₂	NADH
6 x 1.5 = 9 ATP	6 x 2.5 = 15 ATP

Total ATP = 24 ATP

Then, as the acetyl-CoA enters the citric cycle, each pass generates 1 GTP (Same as 1 ATP), 3 NADH and 1 $FADH_2$.

For the citric cycle:

Acetyl-CoA	NADH	$FADH_2$
7 x 1 = 7 GTP	(7 x 3) x 2.5 = 52.5 ATP	7 x 1.5 = 10.5 ATP

Total ATP = 70 from citric cycle.

Total ATP for reaction = Total of oxidation spiral + Total of citric cycle - Activation

$$= 24 \text{ ATP} + 70 \text{ ATP} - 2 \text{ ATP}$$
$$= 92 \text{ ATP}$$

Total ATPs produced = 92

14.75 *a. Name the three ketone bodies.*

Acetoacetate, 3-hydroxybutyrate, and acetone

b. Are they all ketones?

No. 3-Hydroxybutyrate contains no ketone group.

c. Why is acidosis (low blood pH) associated with the overproduction of ketone bodies?

Two of the ketone bodies are carboxylic acids.

14.77 *Where is the NADPH used in fatty acid biosynthesis produced?*

NADPH is produced in the pentose phosphate pathway which occurs in the cytoplasm.

14.79 *Where in a cell does fatty acid biosyntheses take place?*

Cytoplasm

14.83 *During transamination, which molecule is a common amino group acceptor?*

α-Keto acids are amino group acceptors during transamination. A common example is α-ketoglutarate.

14.85 *Oxidative deamination of glutamate produces which compound besides NH_4^+?*

α-Ketoglutarate

14.87 *Which compound is used for nitrogen excretion in the following?*

a. humans

urea in humans

b. birds

uric acid in birds

c. fish

NH_4^+ in fish

14.89 *Why do some scientists believe that mitochondria were once free bacteria?*

Some scientists believe the mitochondria were once free bacteria because they have their own DNA that is different from that of their host cell.

14.91 *How does the way that you generate some of your additional body heat when you are cold differ from the way that an infant does so?*

Adults primarily generate heat by shivering. The heat is produced by the metabolism resulting from muscle action. Newborn infants rely on the oxidation of triglycerides in brown fat cells to produce body heat.

14.93 *a. During digestion, a maltose molecule is hydrolyzed to produce two glucose molecules. Does this reaction take place in the mouth, the stomach, or the small intestine?*

Small intestine.

b. Which enzyme catalyzes the reaction in part a?

Maltase.

c. The two glucoses from part a enter glycolysis and are broken down into pyruvates. Under anaerobic conditions the pyruvates are converted into lactates. Starting from the two glucose molecules and ending at lactates, what is the net change in ATP and NADH?

2 ATP and 0 NADH

d. Under aerobic condition, the pyruvates from part c are converted in acetyl-CoAs, which enter the citric acid cycle. Beginning with two glucose molecules, producing acetyl-CoA's, all of which enter the citric acid cycle, what is the net change in ATP (or its equivalent), NADH, and $FADH_2$?

4 GTP, 12NADH, and 4 $FADH_2$

e. Starting with your answer to part d, and having all of the NADH and $FADH_2$ enter the electron transport chain and oxidative phosphorylation, what will be the net production of ATP, starting with two glucose molecules?

40 ATP

14.95 *Bakers use yeast because the CO_2 produced when it metabolizes sugar causes bread to rise. In yeast, what are the metabolic steps involved in the production of CO_2 from glucose?*

In yeast, glucose is first metabolized to pyruvate through glycolysis. Pyruvate then enters the alcohol fermentation pathway and it is broken down into acetaldehyde and CO_2.

14.97 *One effect of the "low carb" diet craze that swept the United States in 2004 was that a greater number of people had very bad breath, described as "sickeningly sweet." From the standpoint of metabolism, what molecule produced by a diet that is low in carbohydrates and high in fat and protein might cause bad breath?*

A diet low in carbohydrates and high in fat results in the overproduction of acetyl-CoA. Overproduction of acetyl-CoA leads to the formation of ketone bodies, *e.g.* acetone. The release of this product from the body through the mouth could cause a "sickeningly sweet" breath.